Cryopolitics

Cryopolitics

Frozen Life in a Melting World

edited by Joanna Radin and Emma Kowal

The MIT Press
Cambridge, Massachusetts
London, England

This book was set in Stone Sans and Stone Serif by Toppan Best-set Premedia Limited. Printed and bound in the United States of America.

Library of Congress Cataloging-in-Publication Data

Names: Radin, Joanna, editor. | Kowal, Emma, editor.
Title: Cryopolitics : frozen life in a melting world / edited by Joanna Radin and Emma Kowal.
Description: Cambridge, MA : The MIT Press, [2017] | Includes bibliographical references and index.
Identifiers: LCCN 2016031905 | ISBN 9780262035859 (hardcover : alk. paper)
Subjects: | MESH: Cryopreservation | Attitude to Death | Politics
Classification: LCC QH324.9.C7 | NLM QY 95 | DDC 570.75/2--dc23 LC record available at https://lccn.loc.gov/2016031905

10 9 8 7 6 5 4 3 2 1

Contents

Acknowledgments

The authors would especially like to thank Warwick Anderson for intellectual and material support. He, Jenny Reardon, Kim TallBear, Tiffany Romain, and Jenny Brown were members of the 2012 session "Defrost: The Social After-lives of Biological Substance" at the American Anthropological Association and contributed to the earliest articulations of cryopolitics. Along with the contributors to this collection, Marcia Inhorn, Dominique Martin, Alan Petersen, Katharina Schramm, Tim Neale, Ros Bandt, Jennifer Fleming, and Ann-Maree Farrell participated in the Cryopolitics workshop in Melbourne in December 2013 and assisted with our collective thinking on the topic. Karen-Sue Taussig, Alyssa Battistoni, Molly Greene, and Noreen Khawaja provided invaluable insights that helped sharpen our ideas. We are also grateful to Ashley Greenwood, Ava Kofman, Ann Sarnak, and Janneke Koenen, each of whom provided crucial editorial and organizational acumen. Margy Avery first brought this project to the MIT Press, and our editor Katie Helke has done a marvelous job shepherding it to publication. Emma Kowal's work was funded by an Australian Research Council Discovery Early Career Researcher Award (DE120100394) and Discovery Grant (DP150102087).

Freezing Politics

1   Introduction: The Politics of Low Temperature

Joanna Radin and Emma Kowal

In the late 1950s, at the height of the Cold War, a team of scientists participating in the International Geophysical Year traveled to Arctic and Antarctic glaciers (Belanger 2007). This was the beginning of a long-term effort to study the deep history of climate change, buried in snow and concealed in deep layers of ice. This new practice, called "ice coring," transformed the dense, frozen water of glaciers into archives of the atmosphere, providing evidence of recent and significant spikes in greenhouse gases (Alley 2000; Jouzel, Lorius, and Raynaud 2012). Decades, then centuries, and soon millennia were recovered in crystalline cylinders of glacier. Bubbles in these extracted ice cores held a frozen record of seasons past. They contained methane, carbon dioxide, and oxygen isotopes that could be used to track gradual changes in the air.

In the twenty-first century, seventeen kilometers' worth of such ice cores are stored in freezers at the US National Ice Core Laboratory in Denver, Colorado, which boasts one of the largest subzero experimental facilities in the world (fig. 3a.7).[1] Freezing ice cores to study climate change is a practice saturated with ironies. The ability to preserve samples of glacial ice requires energy intensive forms of preservation in order to demonstrate how fossil-fuel-dependent capitalist societies have contributed to climate change. Yet, as the planet warms, "naturally" occurring cold has come to be understood to be a scarce resource. According to NASA, in the summer of 2012, the floating cap of sea ice in the Arctic Ocean "melted to an extent less than half of what it was at the same time of year in 1979, making it one of the most dramatic and visible signs of climate change."[2] Moreover, as rising temperatures thaw the permafrost, long-dormant and potentially pathogenic agents may be released into the atmosphere

(Parkinson et al. 2014). Such pathogens and other forms of cryptic life may be lurking in the freezers of the National Ice Core Laboratory.

How can we make sense of the complex forms of risk and potential produced through these networks of low temperature? Who is responsible for regulating the lives affected by these systems? Are these networks even legible as entities with power over life and death? The archetypical conjunction of temperature, time, and survival involved in the project of extracting and maintaining Arctic ice using energy-intensive artificial freezers—all in in the name of scientific salvation—invites a mode of critical reflection we call *cryopolitical*. This volume, *Cryopolitics: Frozen Life in a Melting World*, presents sixteen essays that give substance to a cryopolitical analytic, drawing on diverse contexts and disciplinary perspectives.

The prefix *cryo*, from the Greek, *krýo* (κρύος), for frost, refers to phenomena that produce cold. In a literal sense, *cryo* signals the forms of stasis made possible through reductions in temperature. As the degrees drop, metabolic processes are slowed and, in some cases, even stopped. Early cryobiologists, the scientists who studied the limits of life at low temperature, called this "latent life" (Keilin 1959). Beginning in the 1930s, Basile Luyet, a Catholic priest and perhaps the first dedicated cryobiologist, sought to understand what life was by examining it from the perspective of what it was not (Luyet and Gehenio 1940).[3] This meant bringing life forms to their limit, a point where all metabolic activity appeared to cease, only to warm them up again and return them to normal functioning. For Luyet and his followers, latent life represented a state of suspended animation in which it was not possible to declare something to be dead or alive. He devoted his career to experimenting with efforts to freeze and thaw different forms of living substance, and his work contributed to the emergence of new forms of artificial insemination, blood transfusion, and tissue banking. These practices of cryopreservation are emblematic of what Alexander Friedrich describes in chapter 3 as a "cryogenic culture."

This cryogenic culture has its roots in the nineteenth century, when the pursuit of cold as a resource for industry, either through the harvesting of ice or its production through new steam-powered technologies, also produced knowledge about the natural world, with social consequences. Chapter 5 presents Rebecca Woods's account of the Victorian era "cold chain" that facilitated a global trade in meat between England and its colonies in the Southern hemisphere. This industrial engagement with low temperature was celebrated in the nineteenth century as allowing the seasons to "shake hands."

In the twenty-first century, forms of preserved substance, both naturally occurring and machine-maintained, comprise our planet's cryosphere, or what Nicola Twilley (2012) has called a "coldscape." The coldscape is an infrastructure, a constellation of social and technical systems that stabilizes otherwise ephemeral and dynamic materials such that they can circulate, producing nutrition, comfort, health, and knowledge, albeit unevenly across the globe. Making and maintaining low temperature, however, comes at a cost.

Cold itself is a matter of degree, and different kinds of machines have been devised to produce and measure different registers of low temperature (Chang 2004). For example, a signal achievement of refrigeration engineering has been to produce cold-storage technologies capable of maintaining very specific low temperatures (Scurlock 1990). It is not uncommon to find laboratories with household refrigerators set to 4°C, ice machines that produce cubes at freezing temperature, upright freezers at –20°C, freezer chests at –80°C (the approximate temperature of dry ice), and liquid nitrogen tanks filled with gas that boils at –196°C. Each of these machines also requires maintenance to ensure its perpetuation as well as that of its contents. Frédéric Keck, in his comparison in chapter 6 of the different temperatures at which samples of influenza are stored and influenza vaccines are stockpiled, argues that these distinct thermal regimes contribute to unscaling human temporality and rescaling it in accordance with shifting ideas of risk that circulate between animal reservoirs and human hosts. The production of these kinds of thermal regimes has also been accompanied by the spaces between them, what we might call the "thermal margins": zones of precarity, ambiguity, and unexpected generativity that also reorganize ideas about what it means to be and to remain alive.[4]

The term "cryopolitics," first coined by Michael Bravo and Gareth Rees (2006; the term is extended by Bravo in chapter 2), was initially intended to draw attention to the emerging geopolitical value of the Arctic in the twenty-first century. Bravo and Rees were interested in how nation-states' efforts to gain control over the valuable mineral resources contained within Arctic territory contributed to climate change without concern for the consequences for Indigenous peoples, flora, and fauna.[5] In this volume, we broaden their concept to encompass the production of artificial low temperature that enables the preservation of human and nonhuman tissues and, in some cases, whole organisms, and even glacial or polar ice itself. Our invocation of "politics" refers more to the tactics and practices that organize and animate science and technology than to formal treaties or doctrines that are often the starting point for inquiries into power.

Cryopolitics, in our conceptualization, intensifies and intervenes in biopolitics, as defined by Michel Foucault. Upon introducing the term in the 1970s, Foucault famously argued that biopolitics pertains to the ways in which power *makes live and lets die* (Foucault 1978). This conceptual axis of biopolitics—with life at one pole, offset by death at the other—has been extremely influential in allowing scholars to examine both life and death as sites at which power acts upon bodies, populations, and territories. Cryopolitics intervenes in this axis of life and death to orient attention to a seemingly paradoxical conjunction of the "cryo" and the "political"—suspended animation and action—that produces a zone of existence where beings are made to live and are *not allowed to die.*[6]

We see an inkling of cryopolitics in Foucault's first airing of biopolitics in the mid-1970s, in a historical context that few scholars of biopolitics have explored. Introducing the concept, Foucault raised a number of then-current examples. He pointed explicitly to the bomb and biotech as augurs of an era of the dubious human ability to intervene more dramatically than ever in the beginnings and ends of life, to "create living matter, to build the monster, and, ultimately, to build viruses that cannot be controlled and that are universally destructive" (Foucault 2003, 253–254). From these intentionally designed forms of making life and death, he suggested, would come problems of politics that transcended national borders and undermined the power of the state to regulate its population. More significantly, these new intensifications in the power to create and destroy life would come to blur the boundaries between life and death.

Along these lines, Foucault also considered how human-built technologies might produce new orientations toward the perpetuation of life and the evasion of death. In his 1976 lecture at the Collège de France, the demise of Spanish dictator Francisco Franco—who had been kept alive in a comatose state for a month before his death the previous year—stood as a symbol of the extension of biopower into the realm of death. "At the moment he himself was dying," Foucault reflected, Franco "entered *this sort of new field of power over life which consists not only in managing life, but in keeping individuals alive after they are dead*" (ibid., xx, emphasis added). The initial articulation of biopolitics, then, emerged as an attempt to come to terms not just with the postwar welfare state and its governance of the living, but also with deathly Cold War technology and its imbrication with politics.[7]

Our interest in cryopolitics is the field of power where the regulation of life is extended indefinitely through technoscientific means such that death appears perpetually deferred.[8] What Foucault situated as an extension of Franco's life through a form of medically enabled life support can also be interpreted as the colonization of life by death. This example is not, in technological terms, a pure example of cryopolitics because it does not rely on the use of low temperature. Yet Foucault's awareness of this vague "sort of new field of power over life" extending into death is consonant with an increasingly broad array of projects that seek to cheat death by freezing life.[9] In the realm of the cryopolitical, life and death both exist to be remade (Franklin and Lock 2003). Take, for example, Ötzi the "Tyrolean Iceman," recovered from a thawing glacier to serve as a spokesperson for human prehistory. As David Turnbull explains in chapter 8, in being subjected to a range of forensic analyses, this long-preserved human cadaver is made to speak about a wide range of issues of human survival, from nutrition to migration. In the process, he has also become legible as a citizen whose body is claimed both by Italy and Austria.

Moreover, a cryopolitical frame asks us to consider the political consequences of the widespread assumption that death is to be avoided at all costs. The need for the concept of cryopolitics, we argue, emanates from the many ways in which low temperatures have increasingly been used to reinforce a tacit devaluation of death. This devaluation has been central to ideas about what it means to be "modern" (see, e.g., Becker 1973; Latour 2013). The concept is perhaps most readily observable in the rarified realm of endangered nonhuman species conservation where biologists see cryopreservation as a safe haven, a refuge, a sanctuary, or a fort, allowing species to dip out of life in order to return at a later time. In Thom van Dooren's account in chapter 13 of species preservation in Hawaii, cryopreservation is presented as a break, a respite from the demands of life in real time. Matthew Chrulew, however, argues in chapter 14 that the project of maintaining such "frozen arks" ultimately eclipses the living animal, attenuating life until it is distilled into a frozen sample that becomes valued only for the genomic information contained within. Eben Kirksey, in his description in chapter 15 of a failed utopia that he and a colleague created for a specific species, the Panamanian golden frog, similarly examines the limitations of aspirations for recovery that congeal around the more ambitious Amphibian Ark project.

This book is motivated by our conviction that in the twenty-first century, life and death can no longer be reduced to discrete phenomenon, nor can either be explained in strictly biological terms. The persistent belief that life—and, for that matter, death—is self-evident does violence to worldviews that configure existence differently. Moreover, such a dichotomous view contributes to the remarkable lack of attention to and administration of *latent life*, the liminal and vague state between life and death. Cryopolitics, in its attention to low temperature and time, pertains to life that is perpetuated indefinitely such that it becomes difficult to assert with confidence that it is not, in fact, death. Put another way, the temporal horizon of cryopolitical life means it is uncertain whether "not letting die" is an alternative technique for "letting die" (cryopolitics as a form of power that *makes live* and *does not let die*) or for "making live" (cryopolitics as power that *does not let die* and *lets die*). Further, it is unclear whether such spaces of ambiguity between life and not-life should be or are already subject to governance. There is a similar absence of agreed-upon techniques for defining and governing cryopolitical life. Take, for example, the case of cryonics, wherein people pay to have themselves frozen, making what Jonny Bunning describes in chapter 11 as "a leap of faith" that a future will arrive in which they will have leapt over death to be thawed into a better world. These individuals put their trust in an as yet unknown form of technoscience that will enable them to be revived, while disregarding an alternative scenario in which their preserved bodies will be regarded by a future society as a source of organs or slave labor.[10]

Cryopolitics, then, is a politics of the liminal. It inheres when life is phantomatic, vacillating, and ambiguous (Squier 2004; Turner 1977; Schrader 2010): when life is not itself (Roosth 2014). If biopolitics in Foucault's formulation referred to the power to make live and let die, cryopolitics is a conceptual intervention that reckons with the status of entities that, for a wide range of reasons, appear to be neither clearly alive nor dead and therefore may be beyond the reach of political power *or* especially vulnerable to it. In cryopolitics, low temperature and indefinite horizons of time combine to create modes of existence defined only by the indeterminacy of their future.

Cold storage has made possible a particular kind of insurance that turns life itself into a source of protection against death. This includes public blood banks as well as private enterprises that provide parents with the opportunity to freeze their children's umbilical-cord blood for potential future use (Waldby 2006; Brown 2012). It also includes seed banks that seek to preserve agricultural possibilities, frozen zoos that maintain gametes and other tissues from endangered animal species and breeds, and IVF clinics that maintain embryos from humans (Fowler, Tefre, and Richardson 2016; Radin 2015; Roberts 2007). Cryopolitical life, by virtue of its ambiguous potential and uncertain horizons of reanimation, often comes to be forgotten or neglected. In the meantime, decision making about when it might ever be appropriate to cease caring for or maintaining frozen materials, or for reviving them, is deferred.[11] As a result, questions about the storage, use, and disposal of frozen material can emerge as unexpectedly heated sites of debate over the sanctity of life itself (Thompson 2005; Radin 2017). This is the most striking temporal dimension of cryopolitics: the abdication of responsibility for action in the present made possible by recourse to the promise of an ever-receding, and technoscientifically enabled, horizon of future salvation.[12]

This promise explains the hold of cold over an increasing array of human and nonhuman life forms. It is a form of "cruel optimism," the product, as Lauren Berlant has argued, of "a system sustained by hopes that can never be fulfilled" (Berlant 2011, cited in Scranton 2015, 23). The cryopolitics of low temperature, of making almost live and not letting die, indexes a form of *cold* optimism, a belief that death—or the acceptance of its finality—can be postponed indefinitely through practices of preservation. The ability to harness low temperature allows for faith in the ability to transport aspects of the present to the future, a future where the good life that eludes us in the present will somehow materialize. But lowering the temperature of optimism does not diminish its cruelty; it merely numbs the pain.

The state of not being allowed to die—of suspending life such that it becomes difficult to discern from death, or extending the process of dying such that it can be mistaken for living—relates not only to political leaders like Franco, to cryonicists, or to nonhuman species.[13] It has come to apply to glaciers and sea ice, the melting of which threatens human survival both directly through rising sea levels and indirectly by endangering life forms that sustain human foodways. In this sense, recent concerns about anthropogenic climate change—often discussed in terms of the "Anthropocene," a proposed geological era that indicts a new degree of human influence on the global environment—warrant interpretation through the lens of cryopolitics.[14] A cryopolitical analysis of the Anthropocene redirects attention away from anxieties about the future to examine the assumptions that guide actions in the present.

By and large, recent efforts to understand climate change have tended to converge on the expenditure of energy and the production of heat (Malm 2015). When did humans begin using fire? When did that use of fire enable industrialization? When did the carbon-based steam power of industrialization and the extractive practices on which it came to depend become eclipsed by the heat produced by burning oil and splitting the atom? Scholars in the humanities and social sciences, such as Dipesh Chakrabarty, have begun to ask, how are we to comprehend humans as a "geological force" that threatens to destroy living ecosystems? How can we reconcile our extraordinary impact on the earth with our seeming inability to fix it (Chakrabarty 2009, 206)? As Bruno Latour puts it in his writings on the Anthropocene, in a reversal of received ideas of history as human action unfolding on an inert planet, "it is human history that has become frozen and natural history that is taking on a frenetic pace" (Latour 2014, 13).[15] In the face of extreme global harms generated by human ingenuity, we "moderns" tend to freeze. Although the desire to deny or defer death has a long history, the cryopolitical impulse has been massively accelerated as many recognize that their modes of production threaten the ability of the planet to support human life.

A cryopolitical approach invites us to reconsider the futures we are actually creating, instead of submitting to the instinct to defer and preserve. At stake is our capacity to be awake to the present, rather than fixated on a future that may never arrive. Klaus Hoeyer, in chapter 10, proposes "suspension" as a concept to explore cryopolitical concerns that stand at the intersection of temporality and temperature. The act of freezing or suspending life in anticipation of future salvation is an impediment to an actually sustainable future brought about through decisive action and accountability to the present. Cryopolitics offers a means of analyzing the deeply rooted and systemic resistance to death and decay that manifests as a refusal

to mourn the demise of the political economic regime of carbon-based capitalism and the philosophical regime that rests on the human as an autonomous agent.[16] The denial that these fundamental Western projects may
already be dead is often managed through practices of freezing (Scranton
2015; Mitchell 2011). In this vein, Deborah Bird Rose's essay on what she
calls the "zone of the incomplete" focuses on the "end time" of environmental crisis, mass extinction, pollution, and climate change that is arguably our contemporary condition. Freezing, among other technologies of
preservation, promises to delay or salvage the outcome of an end time
experienced as extinction. She argues that rather than preventing the end,
freezing merely prolongs it.

Our own interest in cryopolitics began with research into a specific historical case. Beginning in the 1950s, human biologists—biological anthropologists, epidemiologists, and population geneticists—trained their focus
on human communities that they saw as uniquely adapted to the environments in which they lived but destined to disappear. With new access to
technologies of cold storage, including portable refrigerators, dry ice, freezers, and liquid nitrogen, these scientists sought to collect and freeze blood
from members of so-called primitive communities "before it was too late"
(Radin 2013). Scientists hoped that even if these groups dissolved as a result
of migration, admixture, disease, or state-sponsored violence, their preserved blood would persist as a resource for an emerging molecular science
of life.

Those who maintain freezing machines often take for granted that their
contents must not be allowed to perish. Too often, what is stored in biomedicine's freezers outlasts the life cycle of the research program that justified its initial preservation and the scientists who collected it (Couzin-Frankel
2010; Kowal 2013). In the twenty-first century, this blood remained frozen
but indigenous peoples had not vanished. Certain "descendant communities" demanded that their blood be returned to them.[17] Whereas scientists
regarded this preserved blood as a form of latent life, the potential of which
had not been exhausted, some indigenous activists understood it as a form
of incomplete death (Kowal and Radin 2015).

In one high-profile example, members of the Yanomami, an indigenous
group whose territory spans the borders of Brazil and Venezuela, successfully negotiated the repatriation of blood of their community members
that had persisted in North American freezers for over fifty years. As one

spokesperson for the community argued, "Yanomami don't ... take blood to study and later keep [it] in the refrigerator. ... The doctors have already examined this blood; they've already researched this blood. Doctors already took from this blood that which is good—for their children, for the future. ... So we want to take all of this Yanomami blood that's left over" (Borofsky 2005, 64). On one level, this was an indictment of a biomedical research enterprise with no apparent end point. On another level, it was a philosophical observation about the nature of being and the importance of finitude. Warwick Anderson, whose long-standing interests in the politics of exchange have led him to recognize the continuities between the preservation of specimens and the preservation of the colonial archive, reckons in chapter 12 with the ineffable value of such frozen collections, considering that they are not merely corporeal forms of colonial debris but are animated by the specter of that which cannot be frozen: personhood (Stoler 2013).

Cryopolitics turns our attention to what happens when temperature is used to reorient life—including assumptions about what life is—in time. The wager of this volume is that this new critical lens can inspire novel ways to approach survival and even flourishing. Our contributors examine how and why low temperatures have been harnessed in technoscientific attempts to defer individual death through cooling or freezing whole or parts of human bodies (Hoeyer, Bunning); to defer species death through maintaining captive populations of nonhumans in ways that are coextensive with cryopreserved material (van Dooren, Chrulew, Rose, Kirksey); to defer racial death by preserving biospecimens from indigenous people presumed to be facing extinction (TallBear); and to defer large-scale human death through pandemic preparedness (Keck). The ability to deploy artificial cold as a means of deferral or postponement is often articulated by users as a temporal prosthesis, promising it is never too late to become a mother using cryopreserved embryos or to resurrect an individual, race, or species (Radin 2017).

Each of these chapters is inflected by the work of feminist, indigenous, and critical race scholars, all of whom have long sought to reconstitute the relationship between nature and culture as a means of imagining and creating better worlds. Feminist anthropologists of reproduction were among the first to theorize how technology was transforming life and death in the late twentieth century (Franklin and Ragoné 1998; Strathern 1992;

Franklin and Lock 2003). In these accounts, practices of cryopreservation have been silent facilitators of the effort to choreograph ontological states such as "mother," "father," or "parent" to produce forms of biological relation. A cryopolitical mode of reproduction, in which unused gametes and embryos can persist for years on end without ever being thawed, helps us to understand why certain questions about their status are not considered at the time of freezing, but become taken up in debates over the "right to life" when they are thawed. These questions include not only whether or not they are alive and can be used for purposes other than reproduction, but whether nor not maintaining them without any plans for their use or ceasing to maintain them amounts to a kind of death.[18] Freezing is simultaneously an act of creation and of destruction; which one matters more at a given place and time depends on how one is situated (Haraway 1988).

Indigenous thinkers, as well as feminists, have consistently challenged the limitations imposed by the binary between life and death. In chapter 9, Kim TallBear argues for the ascendency of what she calls "an indigenous metaphysic," an insistence that all matter is lively. "We Dakota," she argues, "might say 'alive.' That there is 'common materiality of all that is.' We Dakota might say, 'We are all related.' That agency should be understood as distributed more widely among human and nonhuman beings." TallBear shows how some indigenous perspectives offer a sharp critique—a critique we recognize as "cryopolitical"—of the Western imperative to draw firm boundaries between life, preserved at all cost, and death, avoided at all cost.

In addition to the ethical and ontological diversity this Western view obscures, indigenous, feminist, and postcolonial scholars know too well that efforts to preserve and extend life often apply to certain kinds of life and not others. The nonmolecular animacy that the Dakota recognize is unlikely to be affected by freezing (Radin 2017). Michael Bravo (chapter 2) emphasizes that indigenous peoples living in the Arctic have long regarded ice itself as a form of vitality (see also Wilson 2003). Frozen entities, from biospecimens to glaciers, express movement and animacy according to timescales that may not be universally legible. Cryopolitics helps to diagnose both the Eurocentrism and the anthropocentrism of efforts to conceptualize life at the intersection of temperature and time.

This critique is also found within science itself, particularly in those scientific communities that grapple with the consequences of human-induced environmental change. An early example occurred in 1955 when scientists from the life, earth, and social sciences gathered in Princeton, New Jersey, for a weeklong symposium sponsored by the newly created Wenner-Gren Foundation for Anthropological Research. Their goal was to reckon with evidence that human activity had irrevocably transformed the planet and to determine baselines from which to direct future activities in the service of maintaining life on the planet. The proceedings, published as *Man's Role in Changing the Face of the Earth*, filled more than one thousand pages, capped with a commentary from public intellectual and critic Lewis Mumford (Thomas 1956).

Mumford was suspicious of the values behind the technical language of the discussion. The human, he argued, is a species defined by much more than could ever be captured by scientific abstraction. Furthermore, it was not at all obvious, claimed Mumford, that "technological civilization" itself—the societies from which virtually all of the presenters hailed—should serve as a "baseline" of how to live. There were many possible worlds to be built, he explained, depending on how humanity chose to prioritize what was important. If the only goal was survival, "a uniform type of man, reproducing at a uniform rate, in a uniform environment, kept at a constant temperature" might solve current problems only to create a much deeper one: "why should anyone, even a machine, bother to keep this kind of creature alive?" (ibid., 1151). Mumford's question is fundamentally cryopolitical. In his critical appraisal of the politics of survival, time, and temperature, he implied that there are other, more creative ways to approach human existence. He was invested, as are the authors in this volume, in moving beyond such narrowly conceived conceptions of survival, of nature versus culture, of life versus death, toward more meaningful forms of earthly flourishing.

Description of Chapters

The chapters in this volume—by historians, anthropologists, philosophers, and sociologists of science, medicine, and the environment—examine projects of producing or maintaining low temperatures as means of suspending or postponing processes of decay. They grew out of a workshop held in Melbourne in December 2013 where many of the essays were first presented. Other shorter contributions (by de Chadarevian, Rose, Hoeyer, and

Thompson) address specific ideas that emerged from discussions during and after the workshop. Together, they offer a multidisciplinary treatment of cryopolitics that opens new avenues of inquiry about survival and extinction, preservation and degradation, and other forms of material persistence and loss. They demonstrate how the concept offers a framework for analyzing social and technical strategies for resisting death and decay at multiple scales: the individual, the population, the species, the ecosystem, and the planet.[19] In gaining access to each other's historical and conceptual repertoires, scholars working in the environmental humanities and medical biopolitics can generate fresh insights that transcend divisions between human and nonhuman forms.[20] Each chapter in this collection contributes to charting new pathways for the humanistic and social scientific study of environments—human built, human regulated, and human altered. They reveal submerged connections between the life sciences, environmental sciences, public health, political economy, and technology.

The volume's sixteen chapters are divided thematically into four parts. The first, "Freezing Politics," comprises this introduction, Michael Bravo's "A Cryopolitics to Reclaim Our Frozen Material States," and Alexander Friedrich's "The Rise of Cryopower." Bravo explores the links between the natural "cryosphere"—the poles and high-altitude regions—and the artificial cryosphere—zones of industrial refrigerated and frozen storage and transport. His essay calls for an ecological approach to an empirically grounded cryopolitics that analyzes the flows of "frozen material states." Far from inert, frozen landscapes underpin the life-support systems of the planet and are ecologically, politically, and culturally rich sites of dwelling in their own right. Bravo pushes us to see ice not as obstructive but as flowing, facilitating the lifeways of Northern peoples and, increasingly, the profits of global capital. In chapter 3, Friedrich's attention focuses on the artificial, human-made dimension of the cryosphere. He describes how in the nineteenth century, a "horizontal regime" of artificial coldness was built as a cold supply chain for distributing fresh food (or what he calls *dead life*). In parallel, a "vertical regime" of coldness emerged as a byproduct of physical science. Physicists created highly efficient thermodynamic machines in their race to reach absolute zero, the point where atomic motion effectively stops. After World War II, these new machines were later applied to the horizontal regime for creating and distributing cryobiological entities (*undead life*) using the cold supply chain. Thus a comprehensive infrastructure of artificial coldness, combining the vertical and horizontal regimes, was established by the middle of the twentieth century, constituting the two poles of the cryogenic culture: cryopower and cryogenic life.

This first part of the book concludes with a series of "Freeze Frames," a set of curated images that convey the affective stakes of cryopolitics: sweltering lions cooled by licking blocks of frozen blood, an art installation of human-shaped melting ice statues, and ice cores stockpiled in Denver. These images remind us that cold is as often associated with urgency and crisis as it is with stillness and immobility.

Maintaining the frozen state—be it naturally occurring or artificially produced—requires constant technical intervention. Such interventions might involve the development of machines, the creation of infrastructure, or the engineering of social regimes. The second part, "Technics of Freezing," explores how cold is made and maintained. It begins with Soraya de Chadarevian's meditation on a crucial albeit mundane technology of the lab: the ice bucket. A historian of science, de Chadarevian began her career as a scientist. For her, the ice bucket is an evocative object, a way into the dynamic worlds of life science (Turkle 2007). The simplicity of the technology may make it vulnerable to being stolen or misplaced, but has also been crucial to its enduring success.

Rebecca Woods maintains the theme of the mobility enabled by cold in her account of the nineteenth-century trade in meat. Technologies for producing artificial cold developed in a dynamic relationship between natural philosophy and practical application; the two cannot be disentangled. Together they also produce knowledge about the natural world, with consequences for all kinds of social relationships. Woods's story demonstrates that cryopolitics has included not just ideas and practices related to making live and not letting die, but the reconfiguration of accepted ideas about "the order of things" (Foucault 1971).

This part concludes with Frédéric Keck's account of the production of influenza vaccines, a story told through the frame of cryopolitics. The tale includes the reconstruction of viruses preserved in frozen bodies and the use of frozen biospecimens to produce knowledge that might prevent the spread of bird flu. Keck rejects tropes of messianic redemption (Rose, this vol.) to explain pandemic preparedness, preferring the anthropological vocabulary of hunting and scarcity, storage and stockpiling.

The next part, "Freezing Ontologies," presents three essays that demonstrate the creative potential of cryopolitical analyses. Deborah Bird Rose's essay on what she calls the "zone of the incomplete" interrogates the urge to salvage the world from extinction through practices of preservation. She diagnoses a tendency toward messianic thinking among those who decide which beings will be preserved and which will not. These judges anticipate that, when the end finally arrives, their choices will be vindicated. In her reckoning, the urge to salvage will have found to be mistaken: in the future that more likely awaits us, nothing can be saved from the devastating effects of human-induced environmental destruction. David Turnbull excavates the case of an individual whose experience of being frozen was not a choice. "Ötzi the Iceman" was found preserved in a European glacier in 1991, where he had been for nearly 5,000 years. Turnbull focuses on how the manifold technoscientific efforts to reanimate him have contributed to keeping Ötzi frozen or fixed. Kim TallBear examines the cryopolitical from the vantage point of a different kind of seemingly inanimate material: pipestone. In an essay that brings ideas about the hold of cold into dialogue with recent work on new materialism, TallBear demonstrates how a critical engagement with cryopolitics can productively challenge Western hierarchies of life. Specifically, she is interested in the potential for reconceiving relationships between humans and nonhumans, living and nonliving in view of more just engagements with science. These three essays stake out the theoretical territory that the remaining two sections of the book explore and extend: in short, how freezing exposes and attenuates the porosity of borders between life and death, human and nonhuman, self and other, past and future, animate and inanimate.

"Freezing Bodies," the third part of the volume, comprises three essays on the cryopolitics of human bodies. Beginning with a new medical procedure that freezes trauma patients in order to buy time to repair their wounds, Klaus Hoeyer explores the contradictions of suspension. Suspense can refer to a delay as well as to the heightened emotions of anticipation. Hoeyer traces the connections between temperature and emotions, arguing that the suspense of cryosuspension reflects the ambiguity of freezing process, producing "destabilized definitions and reconfigured relations" of persons and commodities, self and other.

Jonny Bunning, in addition to interpreting the political economy expressed through cryonics, endeavors to historicize the practice, looking at its roots in European and American cultures. Doing so allows him to see cryonics as an instance of "cryopolitics in action," which creates new kinds of social forms that emerge in response to the ways in which freezing troubles kinship. He argues that the "vital arbitrage" undertaken by those who engage in cryonics provides new ways to think about both social and economic relationships and how they evolve over time.

Warwick Anderson draws on the critical theory of Jacques Derrida to meditate on the significance of what he calls the "frozen archive." In this case, he is referring to the human bodily remains accumulated by biomedical scientist and Nobel Laureate Daniel Carleton Gajdusek. A "cold warrior," Gajdusek spent several decades in the middle of the twentieth century traveling to collect blood and other tissues from communities of medical and anthropological significance. Anderson is concerned with the limits of objectivity and concomitant theories of value associated with projects that involve transmuting persons into "immutable" things. He examines how Gajdusek's frozen tissue-based archive presents problems of politics and power that track, but ultimately transcend, those that inhere to the textual, colonial archive.

These essays emphasize the oscillations between subject and object, person and thing, that freezing induces. Bodies and their parts are jumbled by the event of cryopreservation, whether freezing is intended to extend life or to produce a different kind of death. The bodies in these chapters are raced and gendered, illustrating how social power asymmetries are refracted through freezing. Those who Bunning describes as paying to freeze themselves at the point of "death" in the hope of future resurrection are largely white and male. Hoeyer, in contrast, focuses on experimental medical uses of freezing that use predominantly African-American males as poorly informed research subjects.

The scale of the subject/object problematic is refigured in the final part, "Freezing Species," which includes three essays that examine seed and organism banking for the purposes of species preservation. Following as they do the trio of essays on frozen human bodies, what is immediately apparent is the invisibility of individual animals in nonhuman cryopolitical projects. When nonhuman bodies are frozen, the animal is subsumed by the species (see Chrulew), and individual timescales are dwarfed by temporalities of extinction and survival.

Thom van Dooren's contribution focuses on the preservation of certain snail, bird, and plant species in Hawaii, once known as an epicenter of biodiversity, and now known as an epicenter of extinction resulting from widespread habitat destruction and invasive species. The loss of habitat and introduced predators and diseases mean that the hope of reanimation and resocialization of cryobanked species and captive populations is often grim. Nevertheless, van Dooren situates such projects as hopeful sites for honing practices of care for species suspended in the "zone of the incomplete" (Rose, this vol.), hovering on the edge of extinction.

Matthew Chrulew shows how frozen zoo projects developed as an extension of captive breeding in the 1960s. For an increasing number of species, including humans, the ultimate form of species life that is worth saving may be the "informatic species," a virtual body constituted by extracted cellular information for which the optimum temperature is below zero. The "open-ended genome time" enabled by cryopreservation means that extinction of a species or a race, or the exhaustion of individual reproduction, is no longer an end point (Rose, this vol.).

Eben Kirksey describes a project undertaken with artist and biologist collaborators that involved retrofitting a household refrigerator to produce a habitat—a "utopia"—for the Panamanian golden frog. Kirksey and his colleagues sought to extend the salvage project of the Amphibian Ark institutional network into the private spaces cultivated by citizen scientists. But despite their best efforts at negotiating with zoological authorities, no frog ever lived in their fridge. Kirksey views his uninhabited utopia as a "learning ground" to diagnose cryopolitical problems and cultivate "modest hopes" that create possibilities for shared futures.

In her coda to the volume, Charis Thompson provides a short but remarkably sweeping memoir of a life amid ice. From her childhood sipping cold Cokes in the tropics to her adult research on frozen gametes in the context of assisted reproduction, she narrates the affective experience of being repeatedly caught in the hold of cold. Though unique in particulars, her experiences express the manifold dimensions of cryopolitics that have come to shape the experience of being alive, and learning how to die, in a melting world.

In concluding, we return to Bravo's chapter, in which he presents the case of William Scoresby Jr., an experienced whaler who produced one of the first classifications of ice. Bravo relates the story of how Scoresby fashioned a magnifying glass out of ice and used it to concentrate the sun's rays and set a piece of paper on fire, to the "astonishment and delight" of his crew. From the time Europeans began to harness them in pursuit of human industry, ice and cold have been contradictory and surprising substances. Cold chains facilitate nutrition, comfort, health, reproduction, and scientific knowledge even as they draw on energy resources that jeopardize human and nonhuman futures. Melting ice caps index the pathological consequences of technoscience for the earth's climate, while Denver's archived frozen ice cores offer the opportunity to understand, and perhaps reverse, the endangerment of the Arctic and Antarctic. Like Scoresby's icy glass, we hope this volume focuses the energy of science studies scholars inspired by our contributors' efforts to take the raw material of frozen states, across many contexts in space and time, and fashion a theoretical instrument to concentrate our attention on the world and ignite new approaches.

Notes

1. http://www.icecores.org (accessed November 6, 2015).

2. http://climate.nasa.gov/climate_resources/1/ (accessed March 18, 2016).

3. For a discussion of the influence of Luyet on the field of cryobiology, see chapter 1 of Radin 2017.

4. Warwick Anderson (2016) suggests this phrase, inspired by ecologists' attention to edge effects, activity at the borders of different niches.

5. This agenda has been continued by Mia Bennett (2014). See also her blog "Cryopolitics" at https://cryopolitics.com.

6. Cryopolitics is also distinct from what João Biehl (2005) describes as a zone of social abandonment, because of the intense forms of social labor involved with preventing death for those in what Rose (this vol.) refers to as the zone of the incomplete. Warwick Anderson's chapter 12 in this volume describes the convergences and divergences from Derrida's (1994) ideas of the present absences occupied by ghosts and specters.

7. We are grateful to Jonny Bunning for bringing Foucault's discussion of Franco's life extension to our attention.

8. We conceptualize "life" more broadly than the terms set by biological science, in line with Helmreich's (2011) concept of "forms of life."

9. Immortopolitics, or the politics of "eternal life," might be an even more capacious term (Vatter 2010, 2014). Such conceptualizations complicate the focus on killing and death in studies of thanatopolitics or necropolitics. See Agamben 1998; Mbembe 2003.

10. On the historical concept of the "as yet unknown" see Radin 2014. See also Brown, Rappert, and Webster 2000. On the dubious futures facing cryonic bodies, see Doyle 1997; Parry 2004; Farman 2013; Romain 2010.

11. For discussions of the complex temporality of the meantime, see Sharma 2014.

12. On the anthropology of the horizon, see Petryna 2015. Science as a form of salvation has been examined by philosopher Mary Midgley (1992).

13. Or, for that matter, Lenin—whose supporters did try to freeze him, as Bunning describes in chapter 11, this volume.

14. This is the definition of the "Anthropocene" as popularized by geologist Paul J. Crutzen (2002).

15. Of course, the desire to preserve human life and defer the destructive effects of a changing climate is not the only response to the Anthropocene. The most obvious example of an alternative response is climate change deniers, who deal with the threat of climate change by not accepting it.

16. Here we are inspired by work in posthumanist theory that rejects the classic humanist divisions of self and other, mind and body, society and nature, human and animal, organic and technological. See, e.g., Haraway 2008; Hayles 1999; Wolfe 2010.

17. The term "descendant communities" derives from archaeology, but the concept is gaining currency in the management of frozen samples collected from indigenous communities. See, e.g., the National Centre for Indigenous Genomics, http://ncig.anu.edu.au/.

18. This amounts to a recasting of a "biomedical mode of reproduction," a concept developed by Thompson (2005) in the context of assisted reproductive technologies.

19. On the importance of scale in the long history of climate science, see Fleming, Janković, and Coen 2006.

20. This move is in tune with and extends to the realm of the biomedical; see Kirksey and Helmreich 2010; Helmreich 2011.

References

Agamben, G. 1998. *Homo Sacer: Sovereign Power and Bare Life*. Stanford, CA: Stanford University Press.

Alley, R. B. 2000. *The Two-Mile Time Machine: Ice Cores, Abrupt Climate Change, and Our Future*. Princeton, NJ: Princeton University Press.

Anderson, Warwick. 2016. Edge effects in science and medicine. *Western Humanities Review* 69:373–384.

Becker, E. 1973. *The Denial of Death*. New York: Simon & Schuster.

Belanger, D. O. 2007. *Deep Freeze: The United States, the International Geophysical Year, and the Origins of Antarctica's Age of Science*. Boulder: University Press of Colorado.

Bennett, M. 2014. North by northeast: Towards an Asian-Arctic region. *Eurasian Geography and Economics* 55 (1): 71–93.

Berlant, L. 2011. *Cruel Optimism*. Durham, NC: Duke University Press.

Biehl, J. G. 2005. *Vita: Life in a Zone of Social Abandonment*. Berkeley: University of California Press.

Borofsky, R. 2005. *Yanomami: The Fierce Controversy and What We Might Learn from It*. Berkeley: University of California Press.

Bravo, M., and G. Rees. 2006. Cryo-politics: Environmental security and the future of Arctic navigation. *Brown Journal of World Affairs* 8 (1): 205–215.

Brown, N. 2012. Contradictions of value: between use and exchange in cord blood bioeconomy. *Sociology of Health & Illness* 20 (10): 1–16.

Brown, N., B. Rappert, and A. Webster, eds. 2000. *Contested Futures: A Sociology of Prospective Techno-Science*. Aldershot: Ashgate.

Chakrabarty, D. 2009. The climate of history: Four theses. *Critical Inquiry* 35 (2): 197–222.

Chang, H. 2004. *Inventing Temperature: Measurement and Scientific Progress*. Oxford: Oxford University Press.

Couzin-Frankel. 2010. The legacy plan. *Science* 329:135–137.

Crutzen, P. J. 2002. Geology of mankind. *Nature* 415 (6867): 23.

Derrida, J. 1994. *Specters of Marx: The State of the Debt, the Work of Mourning, and the New International*. New York: Routledge.

Doyle, R. 1997. Disciplined by the future: The promising bodies of cryonics. *Science as Culture* 6 (4): 582–616.

Farman, Abou. 2013. Speculative matter: Secular bodies, minds, and persons. *Cultural Anthropology* 28 (4): 737–759.

Fleming, James Rodger, Vladimir Janković, and Deborah R. Coen, eds. 2006. *Intimate Universality: Local and Global Themes in the History of Weather and Climate*, vol. 1. Sagamore Beach, MA: Watson Publishing International.

Foucault, M. 1971. *The Order of Things: An Archaeology of the Human Sciences*. New York: Pantheon Books.

Foucault, M. 1978. *The History of Sexuality*. New York: Pantheon Books.

Foucault, M. 2003. *Society Must Be Defended: Lectures at the College De France, 1975– 1976*. Trans. David Macy. Ed. Arnold I. Davidson. New York: Picador.

Fowler, Cary, Mari Tefre, and Jim Richardson. 2016. *Seeds on Ice: Svalbard and the Global Seed Vault*. Prospecta Press.

Franklin, S., and M. M. Lock. 2003. *Remaking Life and Death: Toward an Anthropology of the Biosciences*. Santa Fe: School of American Research Press.

Franklin, S., and H. Ragoné. 1998. *Reproducing Reproduction: Kinship, Power, and Technological Innovation*. Philadelphia: University of Pennsylvania Press.

Haraway, D. J. 1988. Situated knowledges: The science question in feminism and the privilege of partial perspective. *Feminist Studies* 14:575–599.

Haraway, D. J. 2008. *When Species Meet*. Minneapolis: University of Minnesota Press.

Hayles, N. Katherine. 1999. *How We Became Posthuman: Virtual Bodies in Cybernetics, Literature, and Informatics*. Chicago: University of Chicago Press.

Helmreich, S. 2011. What was life? Answers from three limit biologies. *Critical Inquiry* 37 (4): 671–696.

Jouzel, J., C. Lorius, and D. Raynaud. 2012. *The White Planet: The Evolution and Future of Our Frozen World*. Princeton, NJ: Princeton University Press.

Keilin, D. 1959. The Leeuwenhoek Lecture: The problem of anabiosis or latent life: History and current concept. *Proceedings of the Royal Society of London, Series B: Biological Sciences* 150 (939): 149–191.

Kirksey, S. E., and S. Helmreich. 2010. The emergence of multispecies ethnography. *Cultural Anthropology* 25 (4). 545–576.

Kowal, E. 2013. Orphan DNA: Indigenous samples, ethical biovalue, and postcolonial science. *Social Studies of Science* 43 (5): 577–597.

Kowal, E., and J. Radin. 2015. Indigenous biospecimens and the cryopolitics of frozen life. *Journal of Sociology* 51 (1): 63–80.

Latour, B. 2013. *An Inquiry into Modes of Existence: An Anthropology of the Moderns*. Cambridge, MA: Harvard University Press.

Luyet, B. J., and P. M. Gehenio. 1940. *Life and Death at Low Temperatures*. Normandy, MO: Biodynamica.

Malm, A. 2015. The Anthropocene myth. *Jacobin* (March), https://www.jacobinmag .com/2015/03/anthropocene-capitalism-climate-change/.

Mbembe, A. 2003. Necropolitics. *Public Culture* 15 (1): 11–40.

Midgley, M. 1992. *Science as Salvation: A Modern Myth and Its Meaning*. New York: Routledge.

Mitchell, T. 2011. *Carbon Democracy: Political Power in the Age of Oil*. New York: Verso.

Parkinson, Alan J., Birgitta Evengard, Jan C. Semenza, Nicholas Ogden, Malene L. Borresen, Jim Berner, Michael Brubaker, et al. 2014. Climate change and infectious diseases in the Arctic: Establishment of a circumpolar working group. *International Journal of Circumpolar Health* 73 (10), http://dx.doi.org/10.3402/ijch.v73.25163.

Parry, B. 2004. Technologies of immortality: The brain on ice. *Studies in History and Philosophy of Biological and Biomedical Sciences* 35 (2): 391–413.

Petryna, Adriana. 2015. What is a horizon? Navigating thresholds in climate change uncertainty. In *Modes of Uncertainty: Anthropological Cases*, ed. Paul Rabinow and Samimian Darash, 147–164. Chicago: University of Chicago Press.

Radin, J. 2013. Latent life: Concepts and practices of human tissue preservation in the international biological program. *Social Studies of Science* 43 (4): 483–508.

Radin, J. 2014. Unfolding epidemiological stories: How the who made frozen blood into a flexible resource for the future. *Studies in History and Philosophy of Biological and Biomedical Sciences* 47:62–73.

Radin, Joanna. 2015. Planned hindsight: Vital valuations of frozen tissue at the zoo and the natural history museum. *Journal of Cultural Economy* 8 (3): 361–378.

Radin, J. 2017. *Life on Ice: A History of New Uses for Cold Blood*. Chicago: University of Chicago Press.

Roberts, Elizabeth F. S. 2007. Extra embryos: The ethics of cryopreservation in Ecuador and elsewhere. *American Ethnologist* 34 (1): 181–199.

Romain, T. 2010. Extreme life extension: Investing in cryonics for the long, long term. *Medical Anthropology* 29 (2): 194–215.

Roosth, S. 2014. Life, not itself: Inanimacy and the limits of biology. *Grey Room* 57:56–81.

Schrader, A. 2010. Responding to Pfiesteria piscicida (the fish killer): Phantomic ontologies, indeterminacy, and responsibility in toxic microbiology. *Social Studies of Science* 40 (2): 275–306.

Scranton, R. 2015. *Learning to Die in the Anthropocene*. San Francisco: City Lights Books.

Scurlock, R. G. 1990. A matter of degrees: A brief history of cryogenics. *Cryogenics* 30 (June): 483–500.

Sharma, S. 2014. *In the Meantime: Temporality and Cultural Politics*. Durham, NC: Duke University Press.

Squier, S. M. 2004. *Liminal Lives: Imagining the Human at the Frontiers of Biomedicine*. Durham, NC: Duke University Press.

Stoler, A. L., ed. 2013. *Imperial Debris: On Ruins and Ruination*. Durham, NC: Duke University Press.

Strathern, M. 1992. *Reproducing the Future: Essays on Anthropology, Kinship, and the New Reproductive Technologies*. Manchester: Manchester University Press.

Thomas, W. L. 1956. *Man's Role in Changing the Face of the Earth*. Chicago: Chicago University of Chicago Press.

Thompson, C. 2005. *Making Parents: The Ontological Choreography of Reproductive Technologies. Inside Technology*. Cambridge, MA: MIT Press.

Turkle, S. 2007. *Evocative Objects: Things We Think With*. Cambridge, MA: MIT Press.

Turner, V. 1977. Variations on a theme of liminality. In *Secular Ritual*, ed. Sally Falk Moore and Barbara Myerhoff, 36–52. Amsterdam: Van Gorcum, Assen.

Twilley, Nicola. 2012. The coldscape. *Cabinet* 47, http://www.cabinetmagazine.org/issues/47/twilley.php.

Vatter, M. 2010. Eternal life and biopower. *CR: The New Centennial Review* 10 (3): 217–249.

Vatter, M. 2014. *The Republic of the Living: Biopolitics and the Critique of Civil Society*. Oxford University Press.

Waldby, C. 2006. Umbilical cord blood: From social gift to venture capital. *Biosocieties* 1:55–70.

Wilson, E. 2003. *The Spiritual History of Ice*. New York: Palgrave Macmillan.

Wolfe, C. 2010. *What Is Posthumanism?* Minneapolis: University of Minnesota Press.

2   A Cryopolitics to Reclaim Our Frozen Material States

Michael Bravo

Introduction

The accelerated melting of the cryosphere has been headline news in recent years and has been well documented by scientists. Researchers have extensively studied the significant impacts of the loss of sea ice on shipping routes, habitats, and livelihoods (e.g., Arctic Council 2005, 2009). As well as transforming the surface of the Arctic Ocean, this large-scale melting threatens the habitats of marine mammals such as polar bears and walruses, as well as the food webs they rely on. It also undermines the livelihoods of the Inuit, the traditional "people of the sea ice." It is the most public and prominent harbinger of the "state change" taking place in the Arctic (Young 2009). Sea ice loss is a touchstone for processes reaching far beyond the Arctic, contributing to changes in the climate of the whole earth system. The melting of sea ice and other frozen states such as permafrost adds another dimension to the accelerated warming of the atmosphere caused by greenhouse gases, a picture well established by scientists. However, what hasn't yet been adequately explained are the politics of frozen ecologies, and why they matter for the majority of citizens of the globe living in cities with no special interest in visiting the polar regions. Cryopolitics is the story of how the earth's frozen states have come to matter in the age of the Anthropocene.

To see that ice has political-legal significance, one need only look to its presence in international legal regimes. For example, the United Nations Convention on the Law of the Sea (UNCLOS) has a special sea ice clause (Article 234) that gives states additional rights to regulate shipping where the marine environment has ice-covered waters. Canada, for example, has used this clause to emphasize its regulatory authority over maritime traffic in its High Arctic waters and to emphasize that in practice it actually exercises its sovereignty over those waters (Steinberg et al. 2015, 44–65). Or one might look at the International Maritime Organization's voluntary Polar Shipping Code that sets out safety standards for the operation of vessels in ice-covered waters. Ice is explicitly political in the sense that direct or indirect regulation of it is an expression of political will or power. Internationally sanctioned rules and norms are used to define proper conduct for the care of, and conduct in, frozen coastal and maritime environments.

The central idea behind cryopolitics is a deeper reflection as to why and how the earth's frozen material states have come to be valued, even when they have in the past often been seen as contributing little, at times considered obstructive in their stubborn resistance to life. In the introduction to this volume, Radin and Kowal rightly identify biopolitics as the theoretical context for understanding our potential to hold life in a near-suspended state of animation. The capacity of our earth to host cold ecosystems held still in time through all but the shortest of the polar regions' summer seasons is diminishing all too quickly (Anderson 2009). In this era of anthropogenic global warming, only rarely do we pause to examine the geography of the earth's cooling systems. Our most concentrated or intensive sources of cooling are the polar regions, which house our reserves of frozen states. Ice, permafrost, and the ocean and wind currents of the polar regions are nature's big coolers, receiving truly gigantic flows of thermal energy from the temperate and tropical regions, providing much cooler flows in return. The high-altitude snow and glaciers of the Himalayas, the Andes, and other mountain systems—the so-called Third Pole—also hold considerable reserves of cooling and hold in a stable, crystalline state very large volumes of fresh water that are estimated to supply at least 40 percent of the world's population with their drinking water. These are key buffers that dampen the impact of warming on our global environment: without these, our future climate scenarios are frightening. Though these frozen states are not sufficiently protected by legislation, still they play a vital role in global systems of cooling.

The geography of cooling systems is incomplete without considering the world's major cities and conurbations where refrigerators, freezers, and air conditioners are ubiquitous. These technologies of cooling are crucial for keeping our habitats cool, our food fresh, and our information technologies at a stable operating temperature. These, too, form a crucial part of cryopolitics. Why don't we think of refrigeration as cooling in the same way as we picture ice caps and glaciers? There are two main reasons. Fridges and freezers produce cooling through artifice; the ice caps and glaciers cool because it is in their nature. In other words, they are natural coolers. Although we normally think of the cooling of our homes, office buildings, supermarkets, and transportation networks powered by compressors as though they were fundamentally different from nature, they are indeed connected. To make these connections clearer, the analysis that follows will also explore how cooled states are built into the technologies that play a role in the chilly or frozen space of our urban and polar ecologies.

The reader is invited to imagine the joined-up flow and distribution of fresh foods essential for the survival of urban life as a complex global network of cooled states. Linking the fragile human ecology of cities is a global-scale infrastructure of "cold chains": refrigerated containers traversing the oceans, railways, and roads, connecting producers, distributors, and consumers (Freidberg 2009; Twilley 2014). The circulation of global capital requires these sophisticated systems of precision control cooling and ventilation to maintain our bodies, the food and goods we depend on, to be maintained within a narrow band of temperatures. However, these compressor-driven cooling systems are technologies that in reality can produce cooling only by removing or displacing heat somewhere else. It is paradoxical that the more our economies grow, the more cooling we need to keep our environments temperate, and the more heat, greenhouse gases, and carbon we produce as externalities or unintended by-products. The consequence of being dependent on finely controlled cooling systems on an industrial scale in our temperate cities and the transportation networks feeding them is that we consume ever-increasing supplies of our frozen reserves from the vaults of our poles and mountain regions. My natural science colleagues—the glaciologists, oceanographers, and ecosystems biologists—measure, tabulate, and model the consumption of these frozen reserves, but we haven't yet learned to assign macroeconomic value to these mass-balance models.

The concept of the Anthropocene as an era in which humanity must recognize itself as an integral part of the earth's natural history, and its own role in shaping the earth's future, is an important source of reflection for discussions of cryopolitics. The interconnections between our everyday habitats and the mass consumption of ice made little sense before the story of the industrial revolution. For many centuries, ice in the West was largely perceived as unredeemable nature that placed all life exposed to it in a state of near-death: obstructive to shipping and hostile to life. As Rebecca Woods shows in this volume, the growing consumption of ice was an important aspect of the culture of nineteenth-century industrialization. Ice was sufficiently scarce in temperate regions that it could become a commodity and fetch a good price as a luxury item or a cooling technology, but only very recently has it been imagined as part of the life-support system for our planet, and thus threatened by severe depletion.

Certain groups with interests in resource extraction, notably mining, oil, and gas companies, have sought to capitalize on their knowledge of the diminishing Arctic ice (McKibben 2013). Arctic states, including the so-called new observer states in the Far East, are no less implicated in the drive for returns on investment in Arctic extraction projects. Whether or not this is responsible or just development is a matter of considerable debate. To enable states to be objectively informed about the impacts of a changing climate, the Arctic Council has commissioned forward-looking assessments. The *Arctic Marine Shipping Assessment* (2009), for instance, analyzed the potential demand and usage of Arctic sea routes during the summer navigation season made possible by reductions in the concentration, extent, and thickness of multiyear sea ice (e.g., Smith and Stephenson 2013; Stephenson, Smith, Brigham, and Agnew 2013). The overt critical objections to large-scale industrial-scale resource extraction in the Arctic have come from environmental organizations like Greenpeace and the World Wildlife Fund. They argue that by not putting the brakes on the relatively expensive investments in the Arctic's fossil fuel extraction, the region will become less pristine, and there will be little chance of meeting the two-degree global average temperature warming targets agreed at the COP21 conference in Paris (2016). None of this should be surprising; yet knowing that the global community's energy providers may delay the road to low carbon conversion until it is too late to ensure stable climates and ecosystems is surely a legitimate source of worry. Perhaps strangely, the complex links between the political economy of global-scale industrialization and the needs of mainstream urban societies for a cooled earth on the one hand and the ecosystem dynamics of the living cryosphere on the other are only just beginning to attract more serious attention (e.g., Anderson 2009; Post, Bhatt, Bitz et al. 2013).

Research on the subject of "black carbon" is a good example of science that attempts to link long-term impacts of industrialization to the loss of sea ice.[1] The product of incomplete combustion of fossil fuels from diesel engines and coal-fired power plants (as well as forest fires and other forms of biomass), black carbon has a direct effect on climate warming. In fact it is the "second largest contributor to human-induced climate warming, after carbon dioxide." The International Panel on Climate Change (IPCC) explains that black carbon "is the most strongly light-absorbing component of particulate matter (PM) and has a warming effect by absorbing heat into the atmosphere and reducing the albedo [capacity to reflect solar energy] when deposited on ice or snow." A summary report published by Azzara, Minjares, and Rutherford (2015) for the International Council on Clean Transportation explains that the heavy diesel burned at sea by the international shipping industry accounted for "8 to 13 per cent of all diesel emissions in 2010." The impact on marine black carbon is disproportionately great in the Arctic because 80 percent of ship-based emissions actually take place in the northern hemisphere. The heavily trafficked Great Circle route of the Pacific Ocean and the trans-Atlantic routes cross into high latitudes beyond 40°N. The proximity of this local black carbon to the Arctic means that where deposited on snow or ice, it directly affects the albedo and "can significantly impact climate forcing and ice/snow melt within the Arctic" (Quinn, Stohl, Arneth et al. 2011). Fortunately, there are a number of measures that the shipping industry can take to reduce black carbon emissions; but the industry has been slow to demonstrate the leadership required to push these much-needed changes through (Robertson and Notenboom 2016).

International systems of distribution by shipping, rail, and road are of course an intrinsic part of the global economy. These technologies of distribution are themselves highly reliant on technological systems of artificial cooling to maintain systems in a cooled or frozen state (e.g., Krasner-Khait 2000). Without these technologies, the globalized production and distribution of fresh food could not take place; the contents of our refrigerators would be relatively empty or locally sourced. However, when we eat fresh fruit grown on the other side of the world, we usually overlook the impact of this cold chain of distribution on the supply of cold reserves held in the polar regions. Why is the link between cooled cities and cooled ecological zones counterintuitive? Perhaps we are in the habit of making a distinction between the cryosphere in which frozen states are seemingly produced by nature, and our needs and desires for cooling that are met by artificial technological systems. The distinction begins to break down, however, when, to name just two examples, we recognize the enormous volumes of snow artificially produced in alpine landscapes for recreational purposes (Nöbauer, pers. comm., 2014), or when we acknowledge that industrial cooling systems have an impact on urban climates. The relationship between industrial cooling and natural frozen states may be more than a mere relation—the two may be seen as continuous. Connecting the dots of these seemingly disparate cryological ecologies is a challenging task that makes sense only when we give careful attention to all kinds of *material frozen (or cooled) states* and the agency they possess.

Compelling research in the biological sciences has begun to shed new light on the ecology of life in the cryosphere, drawing attention to the higher-level ecosystem changes caused by marine and coastal sea ice loss (Post et al. 2013). In the environmental humanities, historians (Wilson 2003) and human geographers (Krupnik et al. 2010; Yusoff 2005) have also begun to explore the character of organic agency and its temporalities in frozen states. In keeping with this innovative research, I argue that a concerted effort is required to explore the global ecology of the cryosphere, taking into account the different natures of frozen states—both physically and politically—and the connections between them.

In 2006, Gareth Rees and I introduced the term "cryopolitics" in the *Brown Journal of World Affairs* to argue for a cultural politics of melting polar sea ice (Bravo and Rees 2006). We also drew attention to the increasing visibility of ice. Much political capital was and is invested in the narrative that melting sea ice is the most prominent feature of a "state change" in Arctic politics and ecosystems, and that well-documented sea-ice loss is opening up a new race for resources in the Arctic (Young 2009). We predicted that the Arctic would continue to contribute "new unforgettable names like *Exxon Valdez* and Alaska National Wildlife Refuge (ANWR) to the international language of environmental debates" (Bravo and Rees 2006, 209). In the years since, the term has been taken up in Bennett's (2014) wide-ranging Arctic blog, and in a more limited sense in security studies (Haverluk 2007; Schulman 2009). Developing a robust political ecology is vital if we are to explain how the cryogenic landscapes of the Arctic are becoming much more closely coupled with the high-speed networks of global capitalism. Environmental state changes in the Arctic are taking place at a macroscale ecosystems level, but the danger is that political states are trying to manage this change without seeing a need to change their industrial strategies. What is troubling is that compelling evidence has shown that rather than inspiring reform, these environmental shifts have led to an intensification of older practices because they also present new economic opportunities for resource extraction. This is where I part company with the nation-state-based consensus, which, as many critics have pointed out, is based on short-term interests and calculations of returns (e.g., McKibben 2013).

In the argument that follows, I want to redefine cryopolitics, extending my previous focus on "ice" per se, to take into account the productive agency of natural and artificial "material frozen states" more generally. I identify the dimensions of a new ecological cryopolitics by painting a picture of the ways in which material frozen states are profoundly entangled with local and international political economy. More than that, I want to overturn the long-received view of ice as intrinsically empty of value, indifferent to life, and indeed lifeless. To the contrary, the character of frozen states is that they have a fantastic capacity to be lively and life-giving, holding within them a sense of home, mobility, and temporality stretching from the abrupt change of seasons to organic processes on timescales of centuries and millennia. Far from a view I frequently hear expressed by experts who should know better, the Arctic's seasonally varying ice-covered waters are complex ecosystems that are home to northern societies such as the Inuit, Inuvialuit, and Inupiat, as well as the animal populations and food webs on which their existence depends.

The lifeways of northern indigenous peoples are based on traditional beliefs and social ontologies that have shown remarkable underlying persistence even as they adopt or accommodate significant aspects of market capitalism's culture. Many northern peoples for whom ice is a central material of their environment are also stakeholders who wish to exercise their right to participate in debates and negotiations about mining, oil and gas production, shipping, and other industries, where those activities infringe on or serve their interests. Northern peoples, increasingly dependent on inefficient fossil fuel generators, are implicated in the contradictions of the global economy that sustains the well-being of northern nations but undermines the ecosystems of the high North and contributes to global inequalities.[2] The Arctic is the close neighbor of Northern Hemisphere citizens and its ecologies are indirectly being consumed by the industrialized global economy, and thereby irrevocably altered.

Cryopolitics ought to attend to the historical and cultural preconditions for framing our geopolitical and ecological understanding of frozen states. Mouffe (1999, 2010) has argued that the character of state public policy and its seemingly homogeneous public sphere precludes voice being given to divergent public interests. Hence if public policy debates in practice often turn out to silence irreconcilably different worldviews, then a different political framework that she calls "agonistic pluralism" would acknowledge this divergence and explicitly make space for opposing views (Mouffe 1999, 2010). The need for just this kind of plural understanding of the value of frozen states is a case in point. The widespread failure to acknowledge these preconditions has meant that frozen states circulate in political discourse through a vocabulary that supposes that they are simply solid masses of freshwater and salt water. The principal correlates of these— glaciers and sea ice—are real enough, but so much else is lost in the process about why frozen states ought to matter, to whom, and at what cost. Frozen states, I will argue, are closely connected to systems of global capital, energy exchange, climate, and food production, without which the lives of the urban-dwelling majority who inhabit rich states would be completely different. That being the case, cryopolitics is as much about the social and economic undervaluing of frozen states as it is about forecasting how many ships will sail the Northern Sea Route and at what cost in five, thirty, or fifty years. In fact, these phenomena are closely connected. The question that this chapter begins to address, however provisionally, is how and why.

From Inert Ice to Frozen States

Before I continue, I want to explain what I mean by material frozen states and why they matter. Common sense informed by school chemistry would suggest that freezing and melting, cooling and heating, are almost identical but inverse phenomena. Twentieth-century chemistry has also taught us that the lower states of energy correspond to relative slowness as electrons can move in lower orbits around their nuclei and at reduced speeds. These changes of energy states correspond to the absorption and release of latent heat in condensation and melting. Although it has been customary to think of frozen substances as inert and lacking vitality, the properties associated with freezing have multiple temporalities or periodicities. Snow and ice, like other solid substances, have different kinds of crystal structures. Some are formed very quickly through precipitation. Sea ice, however, can also accumulate over a period of years, its character changing as it grows and ages. Mixtures of ice that can develop over decades can thin significantly over a single season. Glaciers that move with a slow plasticity over centuries can recede very suddenly for reasons that, despite the innovative research I mentioned above on black carbon, still remain poorly understood.

Unlike melting and freezing pure substances in a test tube in a laboratory, these processes are not simply opposite and reversible because, in general, ecological structures and relationships exhibit path dependency. A good analogy is the manufacture of metals such as steel. The properties of a manufactured metal depend not only on the basic ingredients, but the introduction of impurities, the temperature to which the mixtures are heated, and the speed with which it is cooled. The path-dependent cooling of the crystal structure of metals is termed *annealing*—and the enormous range of properties that annealing can produce is paralleled in the properties of the crystal structures of aqueous mixtures and their dependence on environmental conditions (temperature, atmospheric pressure, convection, etc.) for creating the many textures, structures, and dynamics of snow and ice. The thin layer at the top of permafrost can slowly melt and refreeze annually, but when a critical threshold of permafrost melt takes place it collapses and does not regain its structure when refrozen. Similarly, when a dynamic system of sea ice reaches a critical level of melting, it and the life it supports loses the capacity to reform the following season. People who spend a lot of time in these ecologies—hunters, skiers, climbers—are sensitive to these processes because the properties or snow and ice are so important for their ability to know, feel, and move through their environment. However, the complexity of frozen material states and how they are formed has not had much purchase in public discourses about climate, even if extreme winter weather events capture everyone's attention.

Discipline-based thinking in the earth and life sciences has meant that frozen states are often thought of in a restricted way as a branch of glaciology, as though separate from the life systems they help sustain. Although there are shared lessons and overlaps about water-resource conflicts in hydropolitics (Waterbury 1979) and water's importance for homelands and dwelling (of which more below), cryopolitics remains distinctive in terms of the range of material forms to which it pertains. In field ecologies, frozen material states are often heterogeneous mixtures of substances or solutions. Permafrost, for example, varies in its composition of soil, rocks, and freshwater (Williams 1986). The biotic content of soils also varies enormously, so we are not necessarily talking about a uniform substance. *Extremophile* bacteria can live for very long periods in frozen ecological niches (and have attracted interest as intellectual property for commercial applications). Likewise, the mineral (including salt) and bacterial content of water varies considerably, and thus so do its properties.

The extraordinary life of Arctic sea ice has been described by Alun Anderson in *After the Ice* (2009), a study that offers a most thoughtful exposition of the kind of ecological cryopolitics that I hope to encourage. Sea ice, Anderson explains, far from being inert, has an interior of microscopic brine channels full of life that are "no less complex than the interior of a sponge" (Anderson 2009, 150). Lining these tunnels are "algae, often along with dense mats of bacteria." Feeding on these algae are rotifers, which in turn provide food for carnivorous turbellarians and nematodes that move and squeeze through the narrowest of channels. The sudden spring-time spike in algae production "kickstarts" the marine ecosystem, feeding the smallest grazers, amphipods and copepods, which then sustain the fish that are food for larger fish, birds, and marine mammals (ibid.). It is a serious mistake, therefore, to imagine the Arctic's sea ice as simply a world of solid seawater, which historically has been a surprisingly common assumption; we know from marine ecologists that its structure is in part constituted by the feedback systems of the dynamic biodiversity that inhabits it. The consequence of prolonged summer sea-ice loss may be that these ice habitats give way to a rise in open-water plankton abundance. Whether this will produce a rise or a fall in biological productivity is difficult to say, but there is something tragic about watching Arctic summer sea ice disappear without even realizing that it is alive.

Path dependency applies especially to the dynamics of biodiversity. When a woolly mammoth in the melting permafrost is exposed to the atmosphere and begins to decay, clearly there is no going back—though some would argue that by excavating, storing, and removing them on ice, a specimen's blood and DNA can be preserved and perhaps one day cloned, or even bred in captivity, and eventually set free to repopulate the Siberian landscape. In that example, frozen material states play multiple roles in making the mammoth's former habitat, preserving it from decay in death, and creating a place of dwelling for as yet unborn mammoths of a possible future.

Frozen states require work to sustain beyond simple conditions of maintaining temperature and humidity. Just as pastoral landscapes are maintained by the traditional practices of shepherds and their flocks, so too cryological ecologies are the product of specific relations of mobility, feeding, and exchange. Northern peoples of the Arctic do not normally speak of caring for the ice, but ice nevertheless figures practically and symbolically in their ethical understanding of humans' duties and relationships to nonhuman nature.

These examples open up the prospect of taking seriously frozen states as generative and enabling ecologies rather than symbolizing deficiency or absence. If frozen states make the world habitable on different scales— highly localized ecological relationships, regional ecosystems and livelihoods, and as a resource consumed by global thermodynamic cycles—then it makes sense to ask whether there are conservation strategies that can help protect different kinds of frozen states, even if it seems a bit late in the day. For example, creating more marine protected areas could preserve the integrity of cryogenic systems to some extent, even if this were to mean only a slightly reduced rate of sea-ice destruction (Pew Trust n.d.).

The point is that while the earth's frozen states have for long periods operated at *very slow timescales* of thermodynamic compression and movement, these should not be mistaken for signs of *inert* or *barren* states. We know that processes in the cryosphere can also be alarmingly quick and nonlinear, such as the breaking off of polar ice shelves, the retreat of glaciers, the collapse of tundra, and the concomitant impacts on habitats. Cryopolitics is arguably nothing less than a struggle over the temporalities of the globe's frozen states and our growing industrial-scale consumption of them. In the nineteenth century, ice emerged as a commodity that was directly harvested, sold, and consumed. Today, when humanity's capacity to exert increasingly fine levels of control over cooled states is a source of power and profit, the consumption of the earth's frozen materials is now predominantly the indirect and unintended product of hydrocarbon consumption (e.g., black carbon, greenhouse gas emissions). Our blindness to what this destruction of frozen states represents for the planet's future would seem to be part of the growing tide of environmental disasters being experienced across the globe.

Dwelling in Frozen States

Insofar as cryopolitics is a story of complex thermodynamic systems of heat exchange and compression, it is also about recognizing, accounting for, making sense of, and regulating the dynamics of material frozen states. If we are to develop new strategies for the living world to navigate the dangerous waters and ice floes of the Anthropocene, we must understand existing strategies that have given rise to the industrialized production and consumption of frozen objects worldwide. The reasonable fear that the Arctic summer sea ice, glaciers, and ice sheets are being consumed at an accelerated rate, with positive feedback mechanisms coming into play in a way that might suggest these ecosystems are approaching tipping points, should not lead us to forget that other related kinds of frozen states are also deliberately produced and consumed on an industrial scale. Tracking the gradients and material flows between these seemingly disparate phenomena is required to understand how frozen states are valued differentially by those whose dwelling is situated in their midst, and the far greater numbers whose biopolitical relationships with frozen states are mediated by wider infrastructures of markets, distribution systems, and domestic consumption.

This chapter is in effect a call for an extended and integrated analysis of the global economy of material frozen states, by studying their characteristic mobilities (e.g., malleability, viscosity, seasonality), their ecological dynamics and conditions of stability, and their role in maintaining systems of value. What kinds of temporalities determine how frozen states flow? What does it mean to say that these material flows are consumed and transformed into nonfrozen states with different mobilities? What are the consequences when certain technologies of producing and maintaining frozen states are adopted, while others are forgotten, ignored, or avoided? By reflecting on these questions, a distinctive cryopolitics can provide a more informed understanding of Anthropocene dynamics.

Heterogeneous substances, mixtures, solutions, and suspensions move through systems of classification, industrial processes of heating and refrigeration, computer models of global circulation and energy exchange, industrial global logistics systems, indigenous infrastructures including sea ice trails, legal systems to regulate frozen ecosystems, and the urban technological systems that emit heated waste products and exacerbate the accelerated melting of the cryosphere. The complex interactions between these material flows shape and define everyday life in metropolitan centers and the structural inequalities found there, as well as between urban centers and rural environmental regions. It should not be forgotten that many cities have been built in the Arctic, though it contains large ecosystems with diverse frozen states. Furthermore, many kinds of material frozen states constitute recognized objects: cadavers, freezers, frozen human tissues, permafrost, frazzle ice, extremophile bacteria, pancake ice, multiyear sea ice, artificial snow, frostbitten skin, or a carcass of fresh caribou meat.

Work over the last three decades in science studies has taught us a great deal about how objects and states are created and maintained over long-distance networks (e.g., Barry 2013, Latour 1986, 1987, 1993; Law 1986; Harris 1998; Schaffer 2011). As Law (2002, 91) argues, "Objects are the effects of stable arrays and networks," and not simply formed in premolded natural kinds. By acknowledging that frozen material states are contested in fields of unequal power relations, the conditions of these contestations can be traced through the networks and alliances of actors and actants (Latour 1987). Only then can the complex interaction of human and nonhuman agencies within the global cryosphere be recognized, acknowledged, and understood. Following actants through a translation between thoroughly different kinds of frozen material states can open up the ways we think about the polar regions, particularly the construction of hierarchies of scale from the local through the regional to the global.

In proposing an ecological context to cryopolitical networks and systems, I want also to make space for phenomenology and, following Ingold (2000), a relational "dwelling" perspective. The idea of dwelling grounds human experience in such a way as to make us think of the environment as running throughout bodies, within ourselves, constitutive of perception, and never only external to our experience.

This is important for two reasons. First, it opens up an analytical space for understanding our own relationships to frozen states, so that we resist the temptation to suppose that what is frozen is always external to us. Following Law (1987), I argue that ecologies and technologies of frozen states constitute the experience of feeling at home in a healthy body in so many ways and so effectively that we rarely notice or acknowledge them. Second, I see a productive dialogue to be built with other contributors to this volume about the relationship between cryogenic ecologies and the biopolitics of the body. Radin and Kowal diagnose twentieth-century cryopreservation as an ethically complex insistence on deferring death, to "make live and *not* let die" (Radin and Kowal, this vol.). In ecological terms, a grounded cryopolitics is concerned with maintaining the conditions for humans and other species to dwell in a sustainable way, not merely to survive for as long as possible, or to defer our ecological crises for the next generations to deal with.

The changing approaches to gauging a fulfilling life are increasingly bound up in technologies of dwelling centered on architectures of the body, the home, and the city. As such, my argument rests on a materialist politics of frozen states as sustaining. If this seems novel or strange, it is largely because ice has received such bad press in the public imagination, the hallmark of a hostile environment ready to wreak havoc on the unsuspecting, unprepared, or simply unlucky. This imagery is part of an inherited early modern tradition in which the chief significance attributed to ice was that it is an environmental obstacle to navigation, trade, and commerce, simultaneously dead in itself and threatening the well-being of those who sought a passage through it (e.g., Paley 1802). This vision of ice, also closely connected to danger or fear and the aesthetic of the sublime, reached its nadir in the visual and textual imagination associated with Victorian polar exploration (MacLaren 1985; Stafford 1984). There is also a very different materialist tradition in the natural history of the Romantics that restored the affective and generative agency of ice (Wilson 2003), but it is nonetheless the imagery of a substance essentially void of life—fascinating because it spoke to a utopian desire to suspend life by putting its temporalities on hold (Hoeyer, this vol.; Rose, this vol.), a desire that persists today in the popular imagination. These perceptions in differing degrees retain a firm footing among many polar specialists working in politics, international relations, law, governance, the earth sciences, and economics.

Situating Cryopolitics Historically

The image of ice as an intrinsically low-value substance has a long history. The earth's frozen states have received bad press for a very long time—from societies whose agrarian and pastoral values are rooted in systems of value that looked back to the classical world of temperate Mediterranean ecosystems. Formative traditions within the humanities spanning eighteenth-century natural theology to twentieth-century environmental history have significant debts to classical ideas formulated in the context of agriculture and animal husbandry found in temperate climates. We are therefore grappling with a centuries-old tradition of frozen states being morally inflected with associations of being inorganic, barren, desolate, degenerate, forbidding, inert—in short, we have inherited the idea that ice is not only unproductive, but hostile to navigation and commerce: essentially unredeemed. The view that frozen states constitute an absence of natural virtue and an obstacle to be defeated or mastered is found throughout the European visions of empire and navigation.

There is another side to Enlightenment notions of unredeemed landscapes. However wise or misguided, eighteenth-century naturalists saw themselves as providentially duty-bound to put wastelands to productive use by actively intervening in nature to understand and harness its generative capacities. That ice was hostile to navigation and commerce only added to its curious value as a substance for natural philosophers and natural historians to investigate. Take for example John Leslie, who was in turn a professor of natural philosophy and of mathematics at Edinburgh University. Though not the first gentleman of science to explore frozen states (e.g., Robert Boyle carried out experiments on the specific gravity of ice), he carried out many experiments on the nature of heat and frozen states, studying their composition and behavior, in particular on the formation of ice, using dry oats and a bell jar (Leslie 1817). These early philosophical experiments were seen to be applicable to questions of polar climates. John Barrow, the Admiralty's Second Secretary and moving force behind naval exploration, claimed that the suspected breakup of the northern polar-ice barrier was responsible for cold winds that were bringing about the terrible weather causing failed harvests across the northern hemisphere. This resonated with Barrow's *Quarterly Review* readers because 1816 was a year of food scarcity and became known as the year without a summer (Wood 2014). However, Leslie expertly and credibly took apart Barrow's argument using experimental reasoning to demonstrate that a causal link between winds cooled by the Arctic ice and the weather further south was simply not tenable because the heat loss from the ice was insufficient to account for the colder temperatures (Leslie 1818).

One can adduce from Leslie's cryopolitics a number of motivations for his attack on Barrow: Leslie was a religious dissenter and suspected atheist, who followed Hume's views on causation being nothing more than an unvarying sequence of events, and who held a deep dislike for romantic speculation. He saw plainly that Barrow was using very limited evidence of the destruction of polar ice to make the case for a nationalist project to discover and take control of a much-sought-after imperial shipping route. From the Navy's point of view, the loss of sea ice presented an opportunity to project and celebrate British maritime dominance in the Arctic. By exercising naval power it could ensure its future dominance of shipping routes in the face of competition from Russia, which was expanding its fur trade in Alaska and also commissioning voyages of discovery. This cryopolitical nexus of science, religion, nationalism, shipping, and geopolitics was argued very publicly and explicitly. Looking back on this episode two hundred years later, it is uncanny to observe how an argument about the destruction of polar ice could so explicitly serve a set of political interests that were at once domestic (food scarcity) and international (geopolitical maritime power).

While Leslie was carrying out philosophical experiments about heat and cold, he also became a chronicler and reviewer of voyages of polar exploration. Leslie's contemporary, Professor of Natural History Robert Jameson, was building the Edinburgh Museum's collections of Arctic natural history specimens. William Scoresby Jr., an experienced whaler, produced under Jameson's guidance a rich range of studies culminating in his authoritative *Account of the Arctic Regions* (Scoresby 1820). His studies included one of the first classifications of ice, informed by many seasons of field study while whaling (Scoresby 1815). Even his anecdotes were pedagogically rich, as when one day he showed his crew how to carve and polish a magnifying glass out of ice, which he used to focus the sun's rays on a piece of paper and set it alight. This was characteristic of scientific performances in this period when experimental demonstrations producing astonishment and delight were used to pose more subtle philosophical questions: can light pass through transparent ice without heating it, and if so, why? The exceptional richness of observation and understanding found in Scoresby's work earned him a distinguished reputation on Arctic matters, and it would seem from the sales of his two-volume classic study that his readers valued this complexity.

The commercial shipping of ice to warmer climates as a technology of cooling and a source of fresh water, and the storage of ice during winter for summer use, is part of the larger story of refrigeration and freezing technologies beginning in the early nineteenth century (see Woods, this vol.). The frozen goods distribution industry became more profitable when experiments on insulation gave rise to new commercial materials for containers. Tools and techniques were generated for cutting standard-size blocks of ice for "techniques in storage, transportation, and distribution with less waste" (Krasner-Khait 2000). The first of a number of sophisticated and practical systems of refrigerating rail cars was patented in 1867. These played a key role in making viable the new industrial Midwestern cities like Chicago and Kansas City. To satisfy the rapidly growing industrial demand for millions of tons of ice, factories or ice plants sprang up, alongside the harvesting, cutting, and scraping of ice from lakes (ibid.). The impact of these cooling technologies is easy to underestimate. They transformed people's everyday expectations of access to fresh food and what counted as a normal diet. They transformed the landscape and infrastructure of regions of food production and distribution (Cronon 1991). In turn, the very large amounts of energy required for industrial-scale cooling systems for food and transportation were central to the development of modern urbanism. In short, the development of systems for manufacturing and regulating cooled and frozen states has moved from the periphery of the Western geographical imagination to the heart of global thermodynamics in the Anthropocene.

Reclaiming a Progressive Cryopolitics: Materialities, Temporalities, and Generative Capacities

That natural-occurring frozen states can be priced, harvested, distributed, marketed, and sold on a global scale has allowed them to be designated as a kind of natural resource. One of the conditions of a material being marketable is that it has to acquire sociotechnical utility and be meaningful to those it is intended to serve. In that sense, materials are not intrinsically resources; they *become* resources (Zimmermann 1933; cf. Bridge 2009). This was the case in the nineteenth century, when attempts were made to harvest icebergs, cut them into blocks, distribute, and sell them. As refrigeration became cheaper and icebergs became uneconomical to sell, people stopped thinking of them as resources. Today ice has been reinvented as a resource through the novelty of its *genius loci*. Greenlandic ice is melted for its unique water "purity" and is bottled or added to manufactured drinks that are marketed as uniquely "glacial" (e.g., Brooke, *New York Times*, Oct. 18, 2000). A much greater global demand for polar ice may—or may not—be far off. While political leaders and security analysts debate the likelihood of an Arctic resource war, the destruction of the frozen states that are an integral part of Inuit landscapes is already a "resource war" in the sense that these frozen states are being expropriated in what is arguably a violation of what the Inuit leader Siila Watt-Cloutier has called the "right to be cold" and the way of life supported by that coldness (Bravo 2008, 2009).

In a previous paper, "Voices from the Sea Ice" (Bravo 2009), I argued that the narrative linking resources, environment, and risk bears a strong resemblance to development narratives used in Africa and elsewhere, critiqued by scholars like Emery Roe (1995) in *World Development* and James Ferguson (1990) in *The Anti-Politics Machine*. The melting-ice impact narrative framing indigenous people as being "at risk" from climate change has been used by a range of very different political actors to project their interests and to justify their positions with newly formulated Arctic policies. The changing cryosphere documented by scientists has provided a justification for nation-states to reassert their power over Arctic resource peripheries and to reinforce their sovereign rights to continental shelves. This in turn has enabled Arctic states to shift the discursive framing of governance debates away from the highly contested state-indigenous politics of land ownership toward an international consensus predicated on following the rule of law in the UN Convention on the Law of the Sea (1982).

This body of international law is important in many respects, particularly environmental protection, because the Arctic is a very difficult environment in which to carry out marine search and rescue and to clean up oil spills, given its large amounts of ice (Peterson et al. 2003; Wilkinson et al. 2014). Environmental protection legislation serves Arctic states such as Canada and Russia with a means of asserting their respective sovereignty as Asian states such as China, Korea, Japan, and Singapore have come to the Arctic with serious interests in shipping and energy. This is the new geopolitics of the Arctic, in which the cryosphere is being integrated into the calculations of global financial institutions and markets. As with many contemporary economic issues, both state and nonstate actors (e.g., environmental NGOs, indigenous organizations) move where the profit is. Where neo-liberal energy and shipping interests are at work, the dominant narrative acquires a narrow and instrumental character. In this case, "the melting of the sea ice" and accompanying "resource opportunities" have become key terms in the discursive framing of calculations about international investment, risk, and regulation in the Arctic. It remains to be seen in which sectors, at what pace, and in which modes the melting of sea ice translates into large-scale resource extraction.

Cryogenic ecologies *en masse* can be said to fit an expanded definition of a resource by virtue of their ecosystem services (Eicken, Lovecraft, and Druckenmiller 2009; Goodstein et al. 2010). The ambient temperatures of our temperate and tropical latitudes have traditionally been dampened by energy transfers to the polar regions through advective processes in the atmosphere and through thermohaline circulation in the oceans. If global reserves of frozen material states are being depleted to service an unsustainable rate of fossil fuel consumption, one of the main buffers keeping large-scale ecological extinction at bay is endangered.

If there is a shared belief that these frozen material states are scarce and valuable, with political will it should be possible to design conservation strategies and campaigns to protect them. Yet, this will only be plausible if environmentally conscious citizens believe or feel that frozen material states are closely linked to generating or supporting human life. The cost of climate adaptation resulting from the overheating of the globe will increase dramatically this century. Flood barriers against sea-level rise are expected to cost hundreds of billions of dollars; relocation costs for the financial centers and inhabitants situated in future inundated coastal regions are now being estimated by insurers and politicians. Cryological ecologies are at the heart of the thermodynamic predicament of the planet's metabolic system in that resources of cooling (including our reserves of frozen states) are being consumed as rapidly as waste products from burning hydrocarbon reserves are being produced (Bridge 2009, 1222–1223). Cryopolitics is not going to go away any time soon, so strategies are required that acknowledge its centrality in charting a course through the Anthropocene.

Several years ago, I began to wonder why "conservation of ice" was not a more prominent goal among political ecologists and conservationists.

What might a strategy and campaign to "Save the sea ice!" or another kind of frozen state look like? The obvious criticism of this idea is that cooling the planet is hopelessly unrealistic in a world of accelerated warming or that geo-engineering on such a scale would require taking unjustifiable risks (Fleming 2010). And yet, sea ice, glaciers, and ice sheets are melting at an unprecedented rate, and the life systems they sustain are threatened. Conservation organizations invest heavily in conserving charismatic mega-fauna like polar bears, pandas, and elephants, and pursue strategies for conserving important charismatic landscapes like rainforests and wetlands—so why not conserve frozen material states? If the ecological diversity inherent in very cold and frozen states and ecosystems holds limited appeal to donors and conservation supporters, why not frame a campaign to sponsor a retreating glacier or have supporters' names engraved on a beautiful drifting island of sea ice? What resources exist in political ecology to reimagine the generative capacity of some of the most fascinating and wondrous frozen states?

While it is good that species past, like woolly mammoths, and species present, like polar bears, invite audiences to think about frozen habitats over millennia and not just decades, a progressive cryopolitics invites us to rethink how frozen states, however slow and malleable, compressed and unrelenting, or transparent and life-giving, have produced powerful forms of affect. Surely this tells us a great deal about what kinds of things can be protected, remembered, or memorialized. Conserving frozen states is perhaps like conserving temperate habitats in the sense that protecting them also protects other ecosystems and species.

The biggest barrier facing a progressive political ecology of frozen states is the centuries-old denial of the generative capacities of cold states that have been traditionally defined as marginal to the values of temperate agrarian societies. The emergence of the dominant narratives of Enlightenment experiments to improve the productivity of agriculture (Drayton 2000) was predicated on a belief that other kinds of landscapes and soils are unproductive, wasteful, and unredeemed until subjected to agrarian cultivation or pastoral herding practices. The frozen landscapes of the globe's polar regions and mountain ranges, home to the thwarted growth of degenerate species that would be stronger and more fertile in better soils and climates, were assigned marginal values. Naturalists like Linnaeus and Buffon searched for evidence that the Arctic had witnessed better climates in past

times in the hope of one day transforming these peripheries into agricultur-
ally fertile regions (Koerner 1999).

The persistence of this kind of thinking about cryogenic ecosystems is
damaging. Such a profound misunderstanding entails a myopic vision, a
refusal to see and acknowledge the ways in which frozen states (like ice
cores) hold within them elements, patterns, and residues of the collective
memory of the planet: our "frozen histories" (as though the idea were a
contradiction in terms), their seasonal rhythms, and shifting ecologies. The
denial of this space of history creates a vacuum ripe for nostalgia. Here,
then, might be a point of departure for understanding how a sense of dwelling and home is linked to historical yearnings for times past—as well to
forms of desire that insist on a shared forgetting (Connerton 1989).

Ecosystems in frozen states are productive of homelands as well as the
nonhuman food chains that sustain them. In the Arctic, peoples, livelihoods, languages, and ecosystems contain within them the resources of
long memories that are closely tied to the seasonal dynamics of frozen
states. The Inuit of the High Arctic, for example, have traditionally dwelled
on the permafrost-structured tundra for part of the year and moved on to
the seasonally formed sea ice in late winter and early spring to access marine
life at the floe edge. In these seasons, sea ice was home, a place to dwell, a
place to live "on the trail," delineated annually by breaking the same precise network of trails that had melted the previous summer, connecting
named places on the ice marking traditional camps, good hunting areas,
and shared histories (Aporta 2004). Traversing routes across the sea ice and
hunting marine mammals on sea ice and in the adjacent icy waters is still
remembered and enjoyed by many Inuit as a time of plenty, comfortable
living, and full of daylight (Bravo 2008; Brody 2001).

Traditional narratives attached to frozen states by virtue of their central
importance to seasonal life constitute the daily rhythms of navigation and
hunting. The places in which people dwell are sites of remembering and
forgetting. In this environment, frozen states are a language rich in meaning as the tutored eye acquires with study and practice the erudition to read
what is presented in texture, shape, color, and tone. Inuit youth are still
taught to observe the sea ice and learn place names through a process akin
to apprenticeship in which historical narratives play an important part in
offering context, norms, variations, and expectations of what is observed
on any given day (Aporta 2004).

Traditional Inuit knowledge of sea ice and tundra does not simply encode classifications and facts (though these are important). The knowledge is also phenomenological: changes in frozen states (e.g., ice thickness, the feel underfoot, the risks posed by sea ice covered with thin layers of fresh snow), subtle and dramatic, are experienced, weighed, and judged by embodied presence in an environment (Ingold 2000). The experience of dwelling among frozen states speaks to a full spectrum of perceptive registers, full of affect and demanding attentive observation. Mobility across the sea ice has multiple temporalities whereby precise knowledge of the location and movements of animals and the skills required to navigate the sea ice safely are critical for well-being and survival.

If northern peoples have required great knowledge and skill as people of the ice, they arguably face a more formidable struggle in the cryopolitics they confront today. The agrarian ideology of enclosed land tenure that has equated icescapes with wasteland has continued to deny the richness of hunting societies, whose way of life is imbricated in the rhythms of frozen states. I often hear people speaking the language of cost–benefit analysis, saying that the loss of frozen states will be offset by increased marine productivity and transport, bearing fruit for capital investments. This is exactly reminiscent of the appeal to providentialism as legitimacy for the appropriation of others' dwelling spaces in the name of imperialism. The long-standing gendered (and false) characterization of frozen states as degenerate and infertile has followed roughly the same logic as Enlightenment thought that sought, in the guise of redemption (Rose, this vol.), to transform, and in the process destroy, nonconforming ecologies like wetlands and deserts (Drayton 2000).

In reclaiming the term "cyropolitics," I want to think more broadly about other narratives involving ice by imagining material processes of "frozen states" that go beyond the totalizing binary positions of solid/liquid or frozen/melted that underwrite traditional commonsense political arguments about territory or sovereignty. Gerhardt and colleagues have argued that ice is a "liminal substance that combines and confuses properties" of land and ice. Whereas land is traditionally designated as being "amenable to sedenterization and hence territorialization," and water is by contrast perceived to be "resistant to these assertions of control," ice sits uncomfortably; the temporalities of ice formation, movement, and melting hide or complicate ideal notions of permanence so valued by states

(Gerhardt et al. 2010, 1993; Steinberg et al. 2015). The temporal cycles through which sea ice is present and absent are of a different order than land and sea.

In trying to reclaim an ecosystems-based politics of dwelling for frozen states, the distinction between representing "states of nature" and functional "processes" is just as problematic as the land versus water metadistinction. Being frozen is in cultural and political terms normally imagined to be a material state of nature, whereas melting and freezing, though everyday occurrences in nature, are now processes controlled on industrial scales. In nature there is a sense that anthropogenic warming is out of control, in contrast to the medical laboratory where cryogenics epitomizes the fine-tuned control of a state. Taken together, this sequence of binaries and contrasts leads to the odd conclusion that the loss of cryogenic ecologies on a large scale may somehow be offset by the mastery of laboratory and industrial cryogenic technologies on a very small scale. It is as though what we are losing in our control of climate, we are preserving through our technologies of climate control (air conditioning, refrigeration, thermostats). But in the Anthropocene, our global systems of transporting goods across long-distance networks in a regulated temperate, cooled, or frozen state are achieved through the use of thermodynamic compressors or engines that are themselves closely coupled to atmosphere warming and greenhouse-gas emissions. It is therefore not a case of the former being natural and the latter being artificial, as though landscapes were objective material "things" and cryogenic technologies were simply instrumental devices.

In our cities today, states of cooling are acquiring the status of a scarce resource as more intensive dampening systems are required to sustain the conditions that we identify as integral to "being at home." People can learn to be less dependent on air conditioning to keep the outdoor elements at bay, just as they can learn to do without patio heaters. However, the industrial systems of compressor-engines that refrigerate everything from food production to computer servers have added complexity to our political ecology. Our dependence on cooling and freezing technologies to slow down or preserve organic processes is coupled to our high-speed distribution systems. Indeed these cooling systems are a necessary precondition for a significant proportion of global trade, which in turn shapes how our houses are networked through thermostats, WiFi hubs, and cable services, as well as our compressors. How this hardware enables us to feel at home and connected to others so that our level of everyday comfort is confined to a narrow band of temperatures centered on 70°F, neither too hot nor too cold, is the shared condition of contemporary temperate living—within the constraints of inequality and poverty.

Conclusion: Reclaiming a Dynamic Cryopolitics of Home and Dwelling

The energy transfers between the polar regions and the temperate and tropical regions through convection in the atmosphere and oceans are now far better understood than when John Leslie was beginning to model the action of sea ice. The cost of adaptation to the overheating of the globe caused by the engines of capitalism will dramatically increase in the coming decades. My sense is that changing cryogenic ecologies are at the heart of the predicament of the globe. Our cold resources are sinks that we use to offset the heat given off by inefficient combustion. The consumption of these cooled states can come either from our reserves of frozen states (the case with polar ice melting in the face of a warming atmosphere) or from burning more fossil fuels to run refrigeration compressors that create cold states by displacing heat into the atmosphere. Hence our reserves of frozen states are diminishing in response to the changing climate, including both the ongoing interglacial epoch and the anthropogenic component of that change.

A progressive, civic engagement with our industrial way of life increasingly concentrated in large urban centers needs to be at the heart of cryopolitics, not just the strategic interests of powerful states. It requires that we revisit the relationships forged in Enlightenment natural philosophy between naturally frozen states and the artificial cooling of urban centers and global distribution systems. If we were to return to Enlightenment providentialism, we might think we are lucky that the world has large frozen reserves in out-of-the-way regions that have for several thousand years served our planet very well. While it is tempting for our industrial interests to view the polar regions as just another frontier that will be mastered by successive generations of transportation and infrastructure technology, we must understand that frozen states are malleable in how they are conceptualized and used and therefore constitute different kinds of resources with different potentialities. A cryopolitics is needed to examine the many potentialities of malleable frozen states. This realization has not been lost on pharmaceutical industries searching for extremophiles living in frozen soils in which properties of slowness and resistance are economically extremely valuable.

The narratives that reduce complex frozen states to digital binaries, on/off, frozen/melted (the differences in kinds of sea ice can be reduced to simplistic terms of presence and absence), are powerful forms of coding and control. These codes are one means by which frozen states are represented in decision making about investment and risk at the heart of global capital. The frozen systems mediate between the function of cryogenic ecosystems as heat sinks or poses risks to navigation, and the needs of capital to compete for short-term returns on its investments.

These binary representations play a significant role in the computational models that drive governance and economic calculus. This is part of the story of the emergence of modeling and finite state machines developed post–World War II in computational analysis. In these machine systems, the finite states are determinate and have no memory. In computation for industrial and financial markets, models in which key variables are represented in terms of binary states create a platform for certain kinds of rational decision-making models. For example, the presence or absence of ice is a factor in the models for calculating maritime risk and insurance premiums. The models can become more sophisticated, but they adopt a restricted set of criteria to assign values to the geographical distribution, character, and significance of ice. Another example is how corporations and markets discount investment opportunities because of the presence or absence of ice (e.g., drilling for oil in ice-covered waters). International companies are in general more concerned with establishing baselines of environmental data as benchmarks for factoring in risk than they are in understanding the complex generative behavior of ecosystems. There is also a propensity to invoke the future rhetorically, in order to accelerate ecological changes so that they appear to be state changes. Policy makers have been pronouncing the imminent disappearance of the Arctic ice as though it had already happened. This mirrors the tendency in futures markets for states possessing resources to discount risks, or to deregulate, where those risks may constitute serious obstacles to investors. It is no coincidence that Arctic states, multinational investors, and financial markets can be entirely onside with the sea-ice-impact narrative, because the narrative provides a common language and platform for negotiating regulation and risk.

The environmental humanities have an opportunity to reclaim cryopolitics from the narrow confines of multiple instrumental interests by rescuing frozen or cold states from the philosophical gulags created by the advocates of certain models of economic production. Frozen ecological states, for so long demonized in our natural theologies, possess generative agencies that sustain a diverse range of ecologies, near and far. Their positive valences need to be recovered and recognized more clearly. Like all ecologies, sustaining them requires work and maintenance. In that sense, the call to invent a new cryopolitics asks that we show greater appreciation of the importance of our frozen states and how they are created, destroyed, and preserved in their industrial, laboratory, and planetary ecological settings.

Notes

1. The backstory to this line of inquiry is that it was inspired by reconstructions of temperature and precipitation that indicate that glaciers in the Alps should have continued to advance into the twentieth century, whereas by the mid-nineteenth century, paradoxically they began to retreat abruptly (Painter, Flanner, Kaser et al. 2013). Scientists determined that black carbon, first expelled into the atmosphere in abundance by the coal-fueled factories of Europe's and America's industrial revolution, may have contributed to the increasing absorption of solar energy by snow and ice. This absorption is measured by climate scientists as radiative forcing (the difference between sunlight absorbed and energy emitted by the earth). Radiative forcing increased stepwise to "13–17 $W.m^{-2}$ between 1850 and 1880, and to 9–22 $W.m^{-2}$ in the early 1900s, with snowmelt season (April/May/June) forcings reaching greater than 35 $W.m^{-2}$ by the early 1900s" (Painter, Flanner, Kaser et al. 2013). The darkening of the snow caused primarily by emissions from industrialization and forest fires has been shown to decrease the albedo (reflectivity coefficient) of snow and ice, and thereby to increase the radiative forcing, dramatically so in some alpine settings.

2. Throughout this chapter, the term "North" is capitalized only where it is used to refer to a place or region. Hence "northern" is not capitalized.

References

Anderson, Alun M. 2009. *After the Ice: Life, Death, and Geopolitics in the New Arctic.* New York: Smithsonian Books.

Aporta, Claudio. 2004. Routes, trails, and tracks: Trail-breaking among the Inuit of Igloolik. *Études Inuit Studies* 28 (2): 9–38.

Aporta, Claudio. 2009. The trail as home: Inuit and their pan-Arctic network of routes. *Human Ecology* 37 (2): 131–146.

Arctic Council. 2005. *Arctic Climate Impact Assessment*. Cambridge: Cambridge University Press.

Arctic Council. 2009. *Arctic Marine Shipping Assessment 2009 Report*.

Azzara, Alyson, Ray Minjares, and Dan Rutherford. 2015. Needs and opportunities to reduce black carbon emissions from maritime shipping. Working Paper 2015-2, International Council on Clean Transportation.

Barry, Andrew. 2013. *Material Politics: Disputes along the Pipeline*. Wiley-Blackwell.

Barrow, John. 1817. Narrative of A Voyage to Hudson's Bay in His MS Rosamond, containing some Account of the North-eastern Coast of America, and of the Tribes inhabiting that Remote Region. by Lieut. Chappell, R.N. 1817. *Quarterly Review* 18 (35): 431–458.

Bennett, Mia. 2014. Cryopolitics: Arctic News & Analysis (blog). https://cryopolitics.com/.

Bravo, Michael, and W. G. Rees. 2006. Cryo-politics: Environmental security and the future of Arctic navigation. *Brown Journal of World Affairs* 13 (1): 205–215.

Bravo, Michael. 2008. Sea ice mapping: Ontology, mechanics, and human rights at the ice floe edge. In High Places: Cultural Geographies of Mountains and Ice, ed. D. Cosgrove and V. della Dora, 161–176. London: I. B. Tauris.

Bravo, Michael. 2009. Voices from the sea ice: The reception of climate impact narratives. *Journal of Historical Geography* 35 (2): 256–278.

Bravo, Michael. 2010. Epilogue: The humanism of sea ice. In *Siku: Knowing Our Ice: Documenting Inuit Sea-Ice Knowledge and Use*, ed. Igor Kupnik et al., 445–452. Dordrecht: Springer.

Bridge, Gavin. 2009. Material worlds: Natural resources, resource geography, and the material economy. *Geography Compass* 3 (3): 1217–1244.

Brody, Hugh. 2001. *The Other Side of Eden: Hunter-Gatherers, Farmers, and the Shaping of the World*. London: Faber.

Connerton, Paul. 1989. *How Societies Remember*. Cambridge: Cambridge University Press.

Cronon, William. 1991. *Nature's Metropolis: Chicago and the Great West*. New York: W. W. Norton.

Deming, Jody W., and Hajo Eicken. 2007. Life in ice. In *Planets and Life—The Emerging Science of Astrobiology*, ed. J. Baross and W. Sullivan, 292–312. Cambridge: Cambridge University Press.

Drayton, Richard H. 2000. *Nature's Government: Science, Imperial Britain, and the "Improvement" of the World*. New Haven: Yale University Press.

Eicken, Hajo, Amy Lauren Lovecraft, and Matthew L. Druckenmiller. 2009. Sea-ice system services: A framework to help identify and meet information needs relevant for Arctic observing networks. *Arctic* 62 (2): 119–136.

Ferguson, James. 1990. *The Anti-politics Machine: Development, Depoliticization, and Bureaucratic Power in Lesotho*. Cambridge: Cambridge University Press.

Fleming, James Rodger. 2010. *Fixing the Sky: The Checkered History of Weather and Climate Control*. Columbia Studies in International and Global History. New York: Columbia University Press.

Freidberg, Susanne. 2009. *Fresh: A Perishable History*. Cambridge, MA: Belknap Press of Harvard University Press.

Gerhardt, Hannes, Philip E. Steinberg, Jeremy Tasch, Sandra J. Fabiano, and Rob Shields. 2010. Contested sovereignty in a changing Arctic. *Annals of the Association of American Geographers* 100 (4): 992–1002.

Goodstein, E, E. Euskirchen, and H. Huntington, 2010. *An Initial Estimate of the Cost of the Loss of Climate Regulation Services Due to Changes in the Arctic Cryosphere* (unpublished report).

Harris, Steven J. 1998. Long-distance corporations, big sciences, and the geography of knowledge. *Configurations* 6 (2): 269–304.

Haverluk, Terence H. 2007. The age of cryopolitics. *Geography Compass* 50 (3): 1–6.

Ingold, Tim. 2000. *The Perception of the Environment: Essays on Livelihood, Dwelling, and Skill*. London: Routledge.

Koerner, Lisbet. 1999. *Linnaeus: Nature and Nation*. Cambridge, MA: Harvard University Press.

Krasner-Khait, Barbara. 2000. The impact of refrigeration. *History Magazine*. http://www.history-magazine.com/refrig.html (accessed September 26, 2014).

Kristoffersen, B. 2014. "Securing" geography: Framings, logics and strategies in the Norwegian high north. In *Polar Geopolitics? Knowledges, Resources, and Legal Regimes*, ed. K. Dodds and R. Powell, 131–148. Cheltenham: Edward Elgar.

Krupnik, Igor, C. Aporta, S. Gearheard, G. Laidler, and L. Holm. 2010. *SIKU: Knowing Our Ice*. New York: Springer.

Latour, Bruno. 1987. *Science in Action: How to Follow Scientists and Engineers through Society*. Cambridge, MA: Harvard University Press.

Latour, Bruno. 1986. Visualization and cognition: Drawing things together. *Knowledge in Society* 6:1–40.

Latour, Bruno. 1993. *We Have Never Been Modern*. Cambridge, MA: Harvard University Press.

Law, John. 1986. On the methods of long-distance control—vessels, navigation and the Portuguese route to India. In *Power, Action, and Belief: A New Sociology of Knowledge?* ed. John Law, 234–263. London: Routledge.

Law, John. 2002. Objects and spaces. *Theory, Culture, and Society* 19 (5–6): 91–105.

Leslie, John. 1817. New method of freezing water. *Annals of Philosophy* 9 (53): 412.

Leslie, John. 1818. The possibility of approaching the North Pole asserted. … *Edinburgh Review* 30 (59): 193–205.

MacLaren, Ian S. 1985. The aesthetic map of the North, 1845–1859. *Arctic* 38 (2): 89–103.

McKibben, Bill. 2013. *Oil and Honey: The Education of an Unlikely Activist*. New York: St. Martin's Griffin.

Mouffe, Chantal. 1999. Deliberative democracy or agonistic pluralism? *Social Research* 66 (3): 745–748.

Mouffe, Chantal. 2010. *Deliberative Democracy or Agonistic Pluralism*. Political Science Series. Vienna: Institute for Advanced Studies.

Painter, T., M. Flanner, G. Kaser, et al. 2013. End of the Little Ice Age in the Alps forced by industrial black carbon. *Proceedings of the National Academy of Sciences of the United States of America* 110 (38): 15216–15221.

Paley, W. 1802. *Natural Theology or Evidences of the Existence and Attributes of the Deity*. London: J. Faulder.

Peterson, C. H., S. D. Rice, J. W. Short, et al. 2003. Long-term ecosystem response to the *Exxon Valdez* oil spill. *Science* 302 (5653): 2082–2086.

Pew Trust. n.d. Protecting life in the Arctic. http://www.pewtrusts.org/en/projects/protecting-life-in-the-arctic/life-in-the-arctic (last accessed August 19, 2014).

Post, E., U. S. Bhatt, C. M. Bitz, J. F. Brodie, T. L. Fulton, M. Hebblewhite, J. Kerby, S. J. Kutz, I. Stirling, and D. A. Walker. 2013. Ecological consequences of sea-ice decline. *Science* 341 (6145): 519–524. doi:10.1126/Science.1235225.

Quinn, P. K., A. Stohl, A. Arneth, et al. 2011. *The Impact of Black Carbon on Arctic Climate: Report for Arctic Council, AMAP Technical Report No. 4* . http://www.amap.no/documents/doc/the-impac-of-black-carbon-on-arctic-climate/746.

Robertson, Sarah, and Bernice Notenboom. 2016. *Sea Blind*. Montreal: Bear Productions.

Roe, Emery M. 1995. Except-Africa—postscript to a special section on development narratives. *World Development* 23 (6): 1065–1069.

Schaffer, Simon. 2011. Easily cracked scientific instruments in states of disrepair. *Isis* 102 (4): 706–717.

Schulman, Zachary Nathan. 2009. Cryopolitics: The new geopolitics of the Northwest Passage and implications for Canadian sovereignty. MA thesis, George Washington University.

Scoresby, William. 1815. On the Greenland or polar ice. *Journal of the Wernerian Natural History Society* 2 (Pt 2): 261–338.

Scoresby, William. 1820. *An Account of the Arctic Regions*, 2 vols. Edinburgh: Constable.

Smith, L., and S. Stephenson. 2013. New Trans-Arctic shipping routes navigable by midcentury. *Proceedings of the National Academy of Sciences of the United States of America* 110 (13): E1191–E1195.

Stafford, Barbara Maria. 1984. *Voyage into Substance: Art, Science, Nature, and the Illustrated Travel Account*. Chicago: University of Chicago Press.

Steinberg, Philip, Jeremy Tasch, Hannes Gerhardt, et al. 2015. *Contesting the Arctic: Politics and Imaginaries in the Circumpolar North*. London: I. B. Tauris.

Stephenson, R. R., L. C. Smith, L. W. Brigham, and J. A. Agnew. 2013. Projected 21st-century changes to Arctic marine access. *Climatic Change* 118:885–899.

Twilley, Nicola. 2014. What do Chinese dumplings have to do with global warming? *New York Times*, July 25.

Waterbury, John. 1979. *Hydropolitics of the Nile Valley*. Syracuse: Syracuse University Press.

Wilkinson, J., T. Maksym, C. Bassett, A. Lavery, et al. 2014. Experiments on the detection and movement of oil spilled under sea ice. Paper presented at HYDRALAB IV Joint User Meeting, Lisbon.

Williams, P. J. 1986. *Science in a Cold Climate*. Oxford University Press.

Wilson, Eric. 2003. *The Spiritual History of Ice: Romanticism, Science, and the Imagination*. Basingstoke: Palgrave Macmillan.

Wood, Gillen D'Arcy. 2014. *Tambora: The Eruption That Changed the World*. Princeton, NJ: Princeton University Press.

Young, Oran R. 2009. Whither the Arctic? Conflict or cooperation in the circumpolar north. *Polar Record* 45 (1): 73–82.

Yusoff, Kathryn. 2005. Visualizing Antarctica as a place in time from the geological sublime to "real time." *Space and Culture* 8 (4): 381–398.

Zimmermann, Erich W. 1933. *World Resources and Industries: The World Economy*. New York: Harper.

3   The Rise of Cryopower: Biopolitics in the Age of Cryogenic Life

Alexander Friedrich

In classical mythology, Prometheus stole the heavenly fire for the benefit of humankind and enabled the progress of civilization. In the modern era, an equivalent mythic figure might be one associated with the mastery of ice.[1] It is a taken-for-granted reality that modernity is dependent on cooling and freezing technologies. This holds true in a sociotechnical as well as in a bio- and ecotechnical sense. In a sociotechnical sense, modern societies rely entirely on cooling systems that maintain facilities supplying energy, mobility, telecommunication, and much more. Even heat-generating nuclear power plants, coal-fired power stations, combustion engines, electric motors, computers, cellular networks, and the Internet could not function without cooling systems. In a biotechnical sense, cold is used to preserve food, drugs, semen, cells, blood, organs, tissues, and bodies, as well as to condition the environments in which humans live and work. Without cooling devices, there would be a devastating shortage of food, health care, air conditioning, and biomedical services. At the level of ecology, entire cities and even the global climate have become the objects of concerted cooling efforts, as climate change and the heating up of cities induce a growing demand for low temperature.[2]

Cryogenic Culture and Artificial Cryosphere

The large-scale socioecological arrangement that has emerged with the power provided by the ability to harness and control cold has fundamentally shaped modern ways of life. It has also enabled new forms of life due to the achievements of biomedicine and genetic engineering. This set of circumstances amounts to what has been called a *cryogenic culture*.[3] Its formation can be understood through the roughly chronological succession of

(1) the expansion of the ice trade and the invention of mechanical refrigeration in the late nineteenth century (Woods, this vol.), (2) the development of cold chains and their continuing interconnection to global cooling networks, (3) the increasing use of air-conditioning devices in private as well as public buildings and vehicles, (4) the widespread building of cold storage facilities—from central cold stores to domestic refrigerators—as of the middle of the twentieth century, and, finally, (5) the recent formation of cryogenic repositories or cryobanks storing any kind of biomedical substances (Anderson, Chrulew, Hoeyer, all in this vol.). With the integration of all these elements and practices into a large-scale infrastructure, cryogenic cultures can be understood as cultures that both produce and are produced by cold.

Put simply, without constant cooling, a cryogenic culture collapses. The reliability of cooling systems becomes absolutely necessary for social organization. Malfunctions in the cooling infrastructure—ranging from the interruption of communication and transportation services to a large-scale spoilage of organic matter—have the potential to paralyze everyday practices and supply systems. For instance, as historian Jonathan Rees has described, "When Hurricane Katrina struck New Orleans in 2005, the loss of electricity throughout the city ruined refrigerators even in neighborhoods barely affected by the storm as maggots infested the rotting food left behind in them by fleeing residents" (Rees 2013a). When the refugees returned to their houses, they moved the now useless refrigerators onto the streets and sidewalks, hoping that the government would take charge of them. But they stayed there for weeks—silent monuments to the collapse of an infrastructure that is so often taken for granted. This hygienic misery reached truly catastrophic proportions at the central transition points of the cold chain, the interlinked system of technologies that allow low temperature to be maintained across space and time. The New Orleans Cold Storage and Warehouse Company alone operated more than 12,000,000 cubic feet of cold storage space.[4] It took more than four weeks to repair the electricity supply. It then took a small army of decontamination workers six weeks to shovel 26 million pounds of rotten chicken out of the ruined building—almost 12 tons of spoiled meat that subsequently became an opulent feast for rats and maggots (Twilley 2014).

Any cooling system used to keep organic matter fresh can be called *biocentric*. Since the end of the nineteenth century, biocentric cooling technology has enabled a constantly expanding infrastructure that integrates cryogenic cultures into a global topology of *producing, distributing, consuming,* and *disposing of* life as a resource (see figure 3.1). This topology is composed of plantations, slaughterhouses, reefer cargo, cold storages, hospitals, biotech labs, and cryobanks. The manner in which this infrastructure connects to a global space may be called an "artificial cryosphere" (Twilley 2012).

In a geographical sense, the term *cryosphere* denotes any space on planet Earth that is composed of solid water. As a natural space, the cryosphere includes snow, permafrost, pack ice, glaciers, and so on. Depending on season and climatic change, the cryosphere is always growing or shrinking. The natural cryosphere, consisting of water that has been frozen without the use of machines, has become an object of geostrategic interest with the onset of global warming (Bravo and Rees 2006; Haverluk 2007).

The human-built cryosphere, however, suspends the seasons and seems only to expand. In these spaces, an everlasting winter is made to induce artificial hibernation and keep dead organic matter fresh for the sake of the living. As a thermodynamic infrastructure, the artificial cryosphere has had far-reaching consequences for the way life proceeds in modern societies. The cryosphere makes fresh, cooled, or frozen substances available to anyone who is connected to the cold chain. It is the worldwide web of artificial coldness that turns freshness into a technically controlled and highly desired commodity.[5]

Cryopower and Cryogenic Life

As cooled organic matter is the major object of *cryopolitics* in a biopolitical, rather than a geopolitical, sense, this commodity—the actual content of the cryosphere—becomes available as *cryogenic life*. Prevented from aging and spoiling, cryogenic life is the result of the sociotechnical effort to detach organic matter from its natural life cycle. The vast biocentric network of the cryosphere has given birth to a new mode of power—with refrigeration as a main *dispositif* for controlling, transforming, safeguarding, and enhancing life. As a cryogenic modification of biopower, we may call it *cryopower*.

Cryopower is not a mere augmentation of biopower in the Foucauldian sense. Foucault argued that, beginning in the eighteenth century, a new

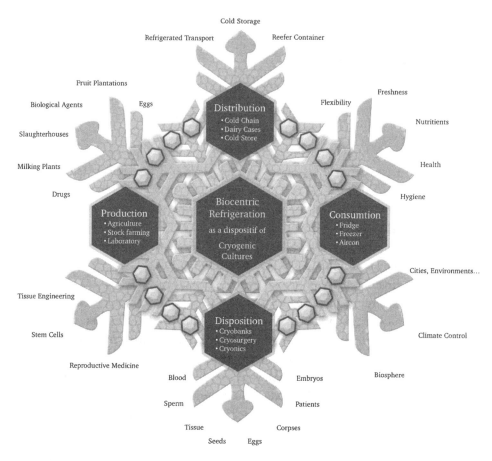

Figure 3.1
The infrastructural basis of cryogenic culture: A topological network of the artificial
cryosphere. Diagram by Alexander Friedrich, designed by Marcel Liebing.

form of power emerged that is not—as is the traditional sovereign power—
based on techniques of disciplining individuals, but on managing populations. In the process, "the basic biological features of the human species
became the object of a political strategy" (Foucault 2009, 1). The "matter of
taking control of life and the biological processes of man-as-species" is what
Foucault (2003, 276–277) calls "biopolitics." As freezing and cooling
technologies became a strategic tool to govern populations, to ensure the
fertility of the species, and to prolong the life of citizens—for example, by
providing fresh food, banked blood, or frozen embryos—these technologies

became the means of the "biopolitics of freezing and cooling" or *cryopolitics*.[6]

As a strategy for connecting transport, security, storage, and administrative spaces to the entire cryosphere, cryopolitics appears to be just the continuation of biopolitics by other—or rather cooler—means. Seen in this way, cryopower, with the cryosphere as its generic *dispositif* is simply an expansion of the biopower described by Foucault (1978a, 134–159). However, taking into account the technospatial infrastructure of the cryosphere and the biopolitical economy of the cooled life that relies on this infrastructure, there are two important differences between Foucauldian and cryogenic biopower. Based on the principle of *make live and let die*, Foucauldian biopower aims to govern human populations as an entity, which is a social mass phenomenon and the biological basis of civil society. Cryopower, however, not only changed the operating principle to *make live and let not die*; it has also changed its primary object. Cryopower is, first of all, no longer limited to human populations, but includes other species as well. Second, it no longer focuses solely on living entities. By keeping organic matter fresh and available as long as possible, cryopower aims to transform *life as such* into *controllable potential*.

As the primary object of this new power, cryogenic life is the cooled source of nutritional value, freshness, health, regeneration, genetic information, new life forms, speculative biocapital, and the standing reserve for purposes "as yet unknown."[7] Cryogenic life as such, however, was not the chief target of biocentric cooling efforts from the beginning. Historically, the primary objects of biocentric cooling technology were products of non-human animal and plant life forms. Originating with the integration of refrigeration devices into transport infrastructures, the initial purpose of the cold chain was the delivery of fresh edibles to feed urban populations in the northwestern hemisphere and social elites in colonial settlements (Tschoeke 1991; Rees 2013b). The development of the cold chain enabled modernizing societies to import perishables from all over the world, allowed urbanized areas and colonial economies to grow rapidly. Once the food supply no longer depended on local, seasonal, climatic, and agricultural constraints, eating habits and consumption patterns changed drastically, thereby opening up new time regimes of labor and social organization (Hellmann 1990; Belasco and Horowitz 2010; Rees 2013b; Gavroglu 2014, 201–280).

Horizontal and Vertical Regimes of Artificial Cold

Dedicated to the preservation and delivery of *dead*, that is, fresh or frozen life, for the sake of the *living*, the cold chain was developed mainly by American engineers and businesspeople for social and economic reasons (Dienel 1991a). Enabling a horizontal distribution of cold and thus launching the creation of the "artificial" cryosphere, the cold chain became a major paradigm of cooling technology. This paradigm can be understood as *a horizontal regime* of artificial cold (Friedrich and Höhne 2014).

A second major paradigm, however, also gave rise to cryopower: the technoscientific achievement of producing the coldest temperatures on earth, and perhaps in the universe. Serving, in the first instance, a purely theoretical purpose, this technoscientific utilization of artificial coldness was a specifically European endeavor dedicated to the race for absolute zero ($-273.15°C$ or $0K$) (Dienel 1991a, 1991b; Gavroglu 2014, 21–132; Mendelssohn 1966; Shachtman 1999). This paradigm may be called the *vertical regime* of artificial cold (Friedrich and Höhne 2014).

Quite early on, though, several technical, personal, and institutional connections emerged between the horizontal and the vertical regime. The research into very low-temperature physics eventually led to the invention of highly efficient thermodynamic machines, vacuum storage vessels, and, with this, the reliable production of liquid gases (Hård 1994). These new technoscientific artifacts became the foundation of a comprehensive economy of biocentric cooling services invented and advanced by the interconnected activities of low-temperature physicists and refrigeration engineers, businesspeople in the globalizing food market, cattle-breeders, scientists, and physicians concerned with the promising prospects of cryobiology—soon recognized and embraced by military officials, politicians, health authorities, eugenicists, and other biopolitical actors after World War II.[8]

As more and more powerful cooling devices invented in the vertical regime were applied to the horizontal regime, the ever-expanding cold chain came to be used not only for food, but also for distributing the semen of breeding animals, for instance, and later blood, mother's milk, and biomedical specimens. In the process, a new cryopolitical regime was established, using the cold chain to generate, maintain, and distribute not only *dead life* for nutritional purposes (fresh food), but *undead life* at very low temperatures (anabiotic or latent life) for the purpose of reviving, refreshing, healing, fostering, and enhancing human or nonhuman populations (Freidberg 2010; Keilin 1959; Landecker 2010). This new, comprehensive regime of biocentric cooling efforts eventually gave birth to both cryopower and cryogenic life as the two poles of the cryogenic culture that continually create and constitutively depend on infrastructures of artificial coldness.

With the linking up of cold chains to a global cryosphere, three developments in the second half of the twentieth century can be regarded as the latest milestones in cryogenic culture: (1) the progress of assisted reproductive technologies with the cryopreservation of gametes; (2) the global trade in cells, tissues, organs, and other cooled, frozen, or vitrified biomedical substances; and (3) the formation of cryobanks in the 1970s, allowing for the storage of millions of human, as well as nonhuman, specimens. With the development of cryobanks, cryogenic life itself eventually became a speculative surplus value, rendering "life" not only a perishable matter, but also a *controllable potential.*[9] This biocentric potential has become the primary object of cryopower and, thus, the strategic main target of cryopolitics.

Cryopolitics as Biopolitical Economy of Cold

The infrastructures maintaining cryogenic life and its potential, however, require huge amounts of energy: the cryosphere, as the global interconnection of cooling devices, is responsible for more than 15 percent of global energy consumption (140 PWh/a) (Gwanpua et al. 2014; Energy Information Administration, http://de.statista.com). This is about three petawatt hours (3×10^{15} kwh), or 68 times as much energy as our planet gives off as heat from the earth's core and mantle. For Nestlé, for instance, the operation of refrigeration devices alone accounts for two-thirds of the food company's total energy consumption (Homsy 2014). Prognoses for the future of energy consumption connected to artificial coldness are quite startling: a recent study of the potential energy demand for cooling in the fifty largest metropolitan areas of the world found that, if air conditioning were to become a social standard in Mumbai, the city alone would have to consume about 24 percent of the demand of the entire United States—while the amount of energy required for air conditioning in the United States already exceeds Africa's total energy consumption (Sivak 2009).[10] The energy demand for all remaining cooling and freezing devices, apart from air conditioning, additionally contributes to the grave disparity in the global consumption of artificial coldness.

Based on these findings, we can conclude that the total energy demand for cooling and freezing technologies economically restricts the growth of the cryosphere and limits global access to it, whereas the public demand for refrigeration and cold storage will continue to increase. Thus, the rise of cryopower and the steady improvement of its icy *dispositif* will continually fan the fires of cryopolitics as the global competition, controversy, and conflict over cooled or frozen resources that conjoin the legal and ethical problems of nourishing and cherishing cryogenic life.[11] The growing urgency of these problems, and the means that will be found to mitigate them, will likely accelerate, rather than impede, the rise of cryopower.

So if there were a mythic figure responsible for having enabled modernity by bringing about coldness, it was a gift that in its crystalline iciness nonetheless stoked the fires of politics, economy, and geopolitical concern.

Notes

1. The prominent version of the Prometheus myth has been given in Plato: *Protagoras*, 320d–322a. For a modern counterpart see Theroux 1982, 35: "Ice is civilization." Cf. Higginson and Smith 1999.

2. See "The Climate Issue" of *National Geographic* (November 2015), which shows planet Earth in a blue marble style on the cover with the title "Cool it."

3. The term *cryogenic culture* was first suggested by Alexander Friedrich und Stefan Höhne (2014).

4. According to information provided by the company's website: http://www.nocs .com/about-us (accessed December 12, 2015).

5. For the history of freshness, especially of food, see Freidberg 2009.

6. See the introduction to this volume and earlier uses of this term in a biopolitical sense in Friedrich and Höhne 2014, and Kowal and Radin 2015.

7. See Radin 2017, esp. chaps. 2 and 5.

8. Friedrich and Höhne (2014) give a more detailed account on the early interconnection between and integration of the horizontal and the vertical regime. See also Radin 2017.

9. For life as a bioeconomical surplus value, see Cooper 2008.

10. Stefan Höhne (2015) gave an instructive talk on this matter at the research colloquium *Space—Place—Power* at Technische Universität Darmstadt (December 1, 2015), where he suggested we should understand the "cryosphere" as an uneven formed "cryoscape."

11. For the latter see, e.g., Waldby and Cooper 2008; Gottweis, Salter, and Waldby 2009; Kowal, Radin, and Reardon 2013; Kowal and Radin 2015.

References

Belasco, Warren, and Roger Horowitz. 2010. *Food Chains: From Farmyard to Shopping Cart*. Philadelphia: Pennsylvania University Press.

Bravo, Michael, and Gareth Rees. 2006. Cyro-politics: Environmental security and the future of Arctic navigation. *Brown Journal of World Affairs* 13:205–215.

Cooper, Melinda. 2008. *Life as Surplus: Biotechnology and Capitalism in the Neoliberal Era*. Seattle: Washington University Press.

Dienel, Hans-Liudger. 1991a. Eis mit Stil: Die Eigenarten deutscher und amerikanischer Kältetechnik. In *Unter Null: Kunsteis, Kälte und Kultur*, ed. Centrum Industriekultur Nürnberg and Münchner Stadtmuseum, 100–111. Munich: C. H. Beck.

Dienel, Hans-Liudger. 1991b. Ganz unten: Vom absoluten Nullpunkt und dem Nutzen tiefer Temperaturen. In *Unter Null: Kunsteis, Kälte und Kultur*, ed. Centrum Industriekultur Nürnberg and Münchner Stadtmuseum, 86–99. Munich: C. H. Beck.

Energy Information Administration. Weltweiter Stromverbrauch in den Jahren 1980 bis 2012 (in Terawattstunden). http://de.statista.com (accessed July 10, 2015).

Foucault, Michel. 2009. *Security, Territory, Population: Lectures at the Collège De France 1977–78*. Lectures at the Collège de France. Basingstoke: Palgrave Macmillan.

Foucault, Michel. Bertani, Mauro, and Alessandro Fontana, eds. 2003. *Society Must Be Defended: Lectures at the Collège De France, 1975–76*. New York: Picador.

Foucault, Michel. 1978a. *The History of Sexuality*, vol. 1: *An Introduction*. New York: Pantheon Books.

Foucault, Michel. 1978b. *The Will to Knowledge* (*The History of Sexuality*, vol. 1). London: Penguin Books.

Freidberg, Susanne. 2009. *Fresh: A Perishable History*. Cambridge, MA: Belknap Press of Harvard University Press.

Friedrich, Alexander, and Stefan Höhne. 2014. Frischeregime: Biopolitik im Zeitalter der kryogenen Kultur. *Glocalism: Journal of Culture, Politics, and Innovation* 1–2:1–44.

Gavroglu, Kostas, ed. 2014. *History of Artificial Cold, Scientific, Technological and Cultural Issues.* Boston Studies in the Philosophy and History of Science, vol. 299. Dordrecht: Springer.

Gottweis, Herbert, Brian Salter, and Catherine Waldby. 2009. *The Global Politics of Human Embryonic Stem Cell Science: Regenerative Medicine in Transition.* Basingstoke: Palgrave Macmillan.

Gwanpua, Sunny George, Pieter Verboven, Tim Brown, Denis Leducq, Bert Verlinden, Judith Evans, Sietze Van der Sluis, et al. 2014. Towards sustainability in cold chains: Development of a quality, energy and environmental assessment tool (QEEAT). In *Refrigeration Science and Technology,* 525–531. London: International Institute of Refrigeration.

Hård, Mikael. 1994. *Machines Are Frozen Spirit: The Scientification of Refrigeration and Brewing in the 19th Century: A Weberian Interpretation.* Frankfurt am Main: Campus.

Haverluk, Terrence W. 2007. The age of cryopolitics. *Focus on Geography* 50 (3): 1–6.

Hellmann, Ullrich. 1990. *Künstliche Kälte: Die Geschichte der Kühlung im Haushalt.* Giessen: Anabas.

Higginson, Ian, and Crosbie Smith. 1999. "A magnified piece of thermodynamics": The Promethean iconography of the refrigerator in Paul Theroux's *The Mosquito Coast. British Journal for the History of Science* 32 (03): 325–342.

Höhne, Stefan. 2015. Kryosphären des Kapitals. Zur urbanen Topologie gekühlten Lebens. Paper presented at Space—Place—Power, Technische Universität Darmstadt, December 1.

Homsy, Paul. 2014. What makes food manufacturing sustainable? Paper presented at 3rd IIR International Conference on Sustainability and the Cold Chain, St Mary's University, London, June 25.

Keilin, David. 1959. The problem of anabiosis or latent life: History and current concept. *Proceedings of the Royal Society B: Biological Sciences,* 150 (939): 149–191.

Kowal, Ema, Joanna Radin, and Jenny Reardon. 2013. Indigenous body parts, mutating temporalities, and the half-lives of postcolonial technoscience. *Social Studies of Science* 43 (4): 465–483.

Kowal, Emma, and Joanna Radin. 2015. Indigenous biospecimen collections and the cryopolitics of frozen life. *Journal of Sociology* 51 (1): 63–80.

Landecker, Hannah. 2010. Living differently in time: Plasticity, temporality and cellular biotechnologies. In *Technologized Images, Technologized Bodies,* ed. Jeanette Edwards, Penelope Harvey and Peter Wade, 211–236. New York: Berghahn Books.

Mendelsohn, Kurt. 1966. *Die Suche nach dem absoluten Nullpunkt.* Kindler.

Plato. 1991. *Protagoras*. Trans. C. C. W. Taylor. Oxford: Clarendon Press.

Radin, Joanna. 2017. *Life on Ice: A History of New Uses for Cold Blood*. Chicago: University of Chicago Press.

Rees, Jonathan. 2013a. The huge chill: Why are American refrigerators so big? *Atlantic*, October 4, 2013. http://www.theatlantic.com/technology/archive/2013/10/the-huge-chill-why-are-american-refrigerators-so-big/280275/.

Rees, Jonathan. 2013b. *Refrigeration Nation. A History of Ice, Appliances, and Enterprise in America*. Studies in Industry and Society. Baltimore: The Johns Hopkins University Press.

Secrétaire Général de l'Association Internationale du Froid, ed. 1908. *Premier Congrès International Du Froid*, vol. 1. Paris: Secrétariat Général de l'Association Internationale du Froid.

Shachtman, Tom. 1999. *Absolute Zero: And the Conquest of Cold*. Boston: Houghton Mifflin.

Sivak, Michael. 2009. Potential energy demand for cooling in the 50 largest metropolitan areas of the world: Implications for developing countries. *Energy Policy* 37 (4): 1382–1384.

Theroux, Paul. 1982. *The Mosquito Coast*. London: Penguin Books.

Tschoeke, Jutta. 1991. Frostige Glieder: Aspekte der Kühlkette. In *Unter Null: Kunsteis, Kälte und Kultur*, ed. Centrum Industriekultur Nürnberg and Münchner Stadtmuseum, 128–141. Munich: C. H. Beck.

Twilley, Nicola. 2012. The coldscape. *Cabinet* 47. http://cabinetmagazine.org/issues/47/twilley.php.

Twilley, Nicola. 2014.. Is your refrigerator running? *Modern Farmer*, January 6. http://modernfarmer.com/2014/01/refrigerator-running/.

Waldby, Catherine, and Melinda Cooper. 2008. The biopolitics of reproduction: Post-Fordist biotechnology and women's clinical labour. *Australian Feminist Studies* 23 (55): 57–73.

Freeze Frames: Life, Time, and Ice

Emma Kowal and Joanna Radin

Figure 3a.1
Lion: A record-breaking heat wave in Australia in January 2014 saw temperatures soar across the arid inland, leading the Bureau of Meteorology to add an extra color-coded category to its official maps for temperatures over 125°F. Coastal cities, where the majority of the population lives, sweltered in up to 113 degrees for five days straight. Scientists and progressive politicians saw it as evidence of the extreme weather patterns exacerbated by climate change. Across the country, people turned to manufactured ice as power failures disabled air-conditioners. Zoo animals were fed iced or frozen meals. Here, Harari, a thirteen-year-old lion, enjoys a blood ice block at Melbourne Zoo. (Source: Rex Features.)

Figure 3a.2

Generation Cryo: Since the first test tube baby was born in 1978, millions of people have been conceived with the help of assisted reproductive technologies (ARTs). Many of these conceptions have relied on frozen donor sperm. Donors have traditionally remained anonymous, with sperm users and their offspring provided with limited information often restricted to height, weight, and eye, skin, and hair color. Fast-forward three decades, and *Generation Cryo* explores the desires of children who want to know where they come from. Premiering in November 2013, this MTV series followed a gaggle of teenagers' encounters with each other as they attempted to find the sperm donor they share as a biological father. These young people situate sperm as central to their search for kinship and identity, illustrating how frozen samples cannot easily be severed from their sources. (Source: MTV.)

Figure 3a.3
Seed Vault: Less than 800 miles from the North Pole, in an abandoned coal mine shaft on the Norwegian Island of Spitsbergen is the Svalbard Global Seed Vault. The Vault—which does not have any permanent staff on-site—officially opened in 2008 and maintains nearly 1.5 million varieties of seeds stored at –18C. They provide a backup for regional seed banks in the event that they fail due to mismanagement, accident, or equipment failure. Sometimes referred to as the "Doomsday Seed Vault," Svalbard has also captured public imagination as a repository that could be accessed to restore agricultural stock following a large scale catastrophe, such as nuclear war. (Source: By Bjoertvedt [Own work], CC BY-SA 3.0, http://creativecommons.org/licenses/by-sa/3.0, via Wikimedia Commons.)

Figure 3a.4
The ice bucket challenge: In the summer of 2014, millions of buckets of ice water were upended on millions of heads. This charity craze raised money for awareness of and research about amyotrophic lateral sclerosis (ALS), the most common form of motor neuron disease. Sufferers experience muscle spasticity and wasting leading to difficulties swallowing, speaking and breathing, usually leading to death within four years of diagnosis. ALS sufferers themselves could not participate in the challenge: in the words of former baseball player Pete Frates, "Ice water and ALS are a bad mix." This did not stop the ice bucket challenge from going viral on social media in late July. Within a month, dozens of celebrities had participated and the US ALS Association received $100 million in extra donations. In this image, Boston City Councilor Tito Jackson and over two hundred other Bostonians take the challenge in Copley Square. (Source: AP via AAP/Elise Amendola.)

Figure 3a.5
The Icemen melteth: Beginning in 2005, Brazilian artist Néle Azevedo began traveling the world to oversee installations of her work. The installations consisted of between 300 and more than 1,000 human-shaped figurines made of ice. Part of a series
called *Minimum Monument*, the figurines were seated in public places, always with the
assistance of people who happened to be there. The anthropomorphic shape of the
frozen water began to melt almost immediately. Those who aided in seating these
"melting men" might have been witnessing their own fates. At the same time, they
may have recognized themselves as members of a polis, capable of evaluating arguments about the fate of humankind. (Source: Minimum Monument in São Paulo-
Brazil, by Néle Azevedo; photo by Fanca Cortez.)

Figure 3a.6
The Iceman: Art often imitates life, but sometimes life imitates art. In 1984, the science fiction movie *Iceman* depicted the encounter between a man who was resuscitated after having been frozen in a block of ice for 40,000 years and an anthropologist who comes to question the morality of the scientists who seek to turn the thawed human into a specimen. The story of the Iceman, nicknamed Charlie, is mirrored in the account of Ötzi, a preserved man discovered in the ice of the Tyrolean Alps in 1991 (the subject of chapter 8). Though Ötzi was not literally reanimated, the various forms of science used to understand him represent an effort to make Ötzi a diplomat from the past to a European cryopolitical present. (Source: Universal Pictures.)

Figure 3a.7
Ice cores: At the height of the Cold War, an international team of scientists sought knowledge about the planet they shared. As members of the International Geophysical Year (IGY, 1957–58), they traveled to Arctic and Antarctic glaciers to begin what would become a long-term effort to study the deep history of atmospheric change. Buried in snow and sealed away in deep layers of ice were bubbles that provided a frozen record of seasons past. These bubbles contained methane, carbon dioxide, and oxygen isotopes that could be used to track gradual changes in the atmosphere. This was the beginning of an effort to mine the earth—not for purposes of exploitation, but for salvation. Decades, then centuries, and soon millennia of environmental changes were recovered. This information was recovered using a new practice of ice coring where cylindrical samples are extracted from deep in the ice. Analysis of these ice cores cast glaciers as silent curators of the atmosphere's archive, eventually providing evidence of significant spikes in greenhouse gases. Once extracted from its natural habitat, the ice core is maintained in a human-made freezer—in this image, at a United States Geological Survey Warehouse—ensuring that it persists as a frozen snapshot of the past. (Source: US Geological Survey National Ice Core Lab, jointly supported by the National Science Foundation.)

Figure 3a.8

Fridgehenge: One of the defining features of Stonehenge is its endurance. Built some-
time between 3000 and 2000 BC, it is thought to have served as an important site of
ritual and as a burial ground. Millennia later, just outside of Santa Fe, New Mexico,
an installation dubbed "Fridgehenge" was erected. It paid homage to the ancient
structure by arranging discarded refrigerators to mirror Stonehenge's ring of standing
stones. Beginning in 1997, Horowitz repurposed machines that ceased to perform
their intended use into a monument of a civilization whose own future has come to
be regarded as endangered. In spite of a hostile municipal council and periodic damage from vandals and windstorms, the artist maintained the installation for a decade
with a team of volunteers. The City of Santa Fe dismantled Fridgehenge in May 2007
after a powerful storm knocked off many of the upper fridges. (Source: Gia Marie
Houck.)

Technics of Freezing

4   Ode to the Ice Bucket

Soraya de Chadarevian

On my first day in the laboratory as an aspiring young biologist, I was given a pristine lab coat. Next, I was shown where the ice buckets were stored. I was handed a square bucket made of white polystyrene and led to an ice machine to fill it up with crushed ice. The ice bucket, I was told, was always going to accompany me when I was preparing an experiment.

Many years later when I had long left lab work behind to become a historian, my scientist interlocutor, running a genomic laboratory, confirmed: "When the ice machine breaks down I am prepared to go very far. Without ice you can't really do anything in biology. The ice bucket allows you to carry your experiment around." In his laboratory the ice buckets were made from polyurethane. They were green and blue and had a lid to prevent the ice from melting too fast. They also had the name of the laboratory written on them in big letters. My interlocutor explained: "Ice buckets are surprisingly expensive and they tend to disappear. You find them floating around in other laboratories."[1]

It is with this apparently indispensable and yet humble technology of the ice bucket that I will be concerned here. On its own it looks so low-tech that even calling it a technology may seem presumptuous. Yet on closer inspection it quickly becomes clear that the ice bucket is a central link in an elaborate cold chain that sustains the study of life in the laboratory.

When American photographer Catherine Wagner spent an artist residency in various genomic laboratories, she noted the freezers. A series of sharp black-and-white photographs titled "–86 degree freezers (twelve areas of concern and crisis)" gives us a look into these spaces of cold storage, opening up the view on semi-ordered stacks of boxes, piled-up plates and racks of test tubes inscribed with letters and figures and covered by a frosty

Figure 4.1
Polyurethane ice buckets with lid. Photograph by the author.

layer inside the freezer. Nothing seems to be moving in these pictures (Homburg 1996).

Entering any biological wet lab today, the visitor would notice the freezers, which are closely integrated with laboratory routines. Test tubes, petri dishes, reagents, samples, and mutants are taken out of the freezer or placed back into it. The laboratory is likely to be equipped with a series of freezers, rather than just one, sometimes even a walk-in freezer or cold room. The laboratory that I visited most recently was fitted with a large –80 degree Celsius freezer, a –20 degree one, a +4 degree fridge and an extra-cold liquid nitrogen tank that keeps biological samples at about –196 degrees (the boiling point for liquid nitrogen), a temperature at which virtually all biological activity stops. The freezers were lined up along a wall facing the lab benches. An ice machine that supplies the ice for buckets was located in a small room across the corridor and shared with other laboratories.

With the exception of the liquid nitrogen freezer, the fridges and ice machines look much like the ones we find in domestic surroundings. They are indeed often produced by the same manufacturers, such as Frigidaire, which entered the refrigerating business in the early twentieth century and whose brand name became so well known that it became a generic term for refrigerators. For many scientists and observers, these machines might be so mundane as to be virtually invisible. Yet familiarity with the cooling equipment and its ubiquity should not deceive us. For David Edgerton, refrigerators, together with sewing machines, spinning wheels, condoms, bicycles, corrugated iron, and chain saws, belong to a category of pervasive "technologies in use," invisible to a history of technology exclusively fixed on innovation but essential in the lives and work of many people around the world (Edgerton 2006, xii).

While their utility to preserve biological material was already well established in the laboratory, only since the mid-twentieth century has the development of new protocols that allowed cells, sperm, and embryos to be frozen and thawed again while preserving their vital functions meant that freezing processes became an integral part in the way life was managed and studied in the laboratory. Freezing and thawing took the place of the much more laborious work of stock maintenance. Whereas samples might previously have been dried, pickled, fixed, and sliced to be preserved and studied, they could now be kept in "animated suspension" (Landecker 2007). Blood, too, could be collected and preserved on ice for future investigations, collapsing and jumbling biological timescales (Radin 2013).

It is in this new experimental regime that the ice bucket finds its place. While freezers occupy a fixed place in the laboratory, ice buckets allow frozen things to be moved around—from the fridge, freezer, or preparation room to the bench or even outside of the laboratory. Moving with the experimenter while she collects and prepares material and follows various steps of the experimental procedure, the ice bucket is a "safe companion" that keeps tissue samples, cells, fractionation products, pellets, enzymes, and other reagents cold, preventing degradation and other unwanted biological processes from taking place while the important work of experiment is underway.[2] The instruction "keep on ice" appears as a refrain in endless experimental protocols.

Reflecting on the function of the ice bucket more carefully, we can distinguish two different ways in which it works, depending on the previous state in which the samples or reagents were kept. If the material comes from a freezer, the ice bucket provides a milieu for gently thawing it. Once thawed, the material can be handled while the ice protects it from degrading too rapidly. The process is unidirectional: once thawed, the samples cannot be frozen again. Alternatively, the ice bucket can help preserve living material or freshly prepared tissue samples by keeping them cold and thus dramatically slowing down all biochemical reactions. It is indeed curious that to observe life processes, we often need to arrest them.

"In biology you really only get snapshots of living processes. You need to freeze them in time to observe them," my interlocutor reminded me, drawing a connection between the technical and epistemic conditions under which life can be studied in the laboratory. The ice bucket may also play a more integral function in experimental designs. Biochemical processes, such as enzymatic reactions, are temperature dependent. Paired with a regulated water bath, the ice bucket becomes part of a temperature-control system that makes it possible to stop, start, and study biological reactions under controlled conditions.

The centrality of the modest ice bucket in laboratory routines becomes most apparent when things go wrong. To perform its function, the ice bucket depends on the ready supply of crushed ice produced by an ice machine. The manual for the Hoshizaki ice maker, a widely used brand, reveals that the machine is a more complex piece of equipment than we might initially appreciate. The machine has to be properly hooked up to the water and drainage system as well as to the electrical grid. Hot gas needs to have an outlet. Only qualified service technicians should install, service, and maintain the ice maker. Likewise, owners should not proceed to operate the ice maker until the installer has instructed them on its proper operation. The 200-page thick *Hoshizaki Technician's Pocket Guide* indicates that there are many ways in which the machine can fail.[3]

Ice buckets, too, have their pitfalls. The ice not only keeps the specimens cool but also helps to keep them in physical place. When the ice is not replaced frequently enough and starts melting, anything placed on the ice loses its moorings. A dreaded scenario is that test tubes or beakers reverse their content into the icy water, rendering an experiment worthless.

To overcome these challenges, manufacturers have developed and advertised benchtop coolers, molded from durable polycarbonate and filled with a nontoxic insulating gel, as an alternative to the "messy ice buckets." According to one recent manufacturer, the lids feature a gridded write-on surface for sample identification and the coolers are stackable, saving "valuable lab space."[4] Also on the market are various kinds of "chill buckets" filled with cooling beads. Advertisements claim that they "re-invent" the ice bucket by doing away with "melted ice, puddles, floating tubes, or wet and contaminated samples." Some of them are molded so as to be conveniently "balanced in one arm," allowing scientists or technicians to freely use one arm while carrying precious cooled research material with another.[5] But so far none of these innovations has managed to supplant the ice-filled buckets. Most scientists still prefer traditional ice buckets, not least because of their more modest price tag. Two suppliers, Fisher and VWR, dominate the American market. In their publicity materials they highlight the versatility, durability, and high performance of their polyurethane-skinned foam buckets. These improved versions of the original ice bucket guarantee "long-term thermal storage of ice and dry ice on the lab bench." They will not "sweat or leak" and are "unbreakable." They are "stable when full." Molded handles enable "safe handling and transport" and a flush lid allows "stacking for convenient storage." Some even have special attachments for pipettes or to keep racks of tubes in place in the bucket.[6]

Seemingly low-tech ice buckets persist alongside high-tech scientific apparatuses in today's labs. They help maintain the cold chain, prevent degradation, control biochemical processes, and make biological materials mobile. What initially looked to be so mundane an object that it could be overlooked now appears as the quintessential piece of equipment of any biological wet lab. No wonder it is keenly guarded.

Acknowledgments

I thank Patrick Allard (UCLA) for his crucial role in this project and Joanna Radin, Emma Kowal, and Warwick Anderson for inviting me to participate at the Cryopolitics workshop in Melbourne in December 2013, where the idea for this piece originated.

Notes

1. Patrick Allard, personal communication, April 14, 2014.

2. Patrick Allard, personal communication, April 14, 2014.

3. Hoshizaki America Inc., Modular Flaker, Models F-801MAH (-C) and F-801WH(-C): Instruction Manuel (issued 2003; revised 2010); http://www.hoshizaki.com/docs/manuals/F-801M_H(-C)_serv.pdf and Tech-Spec's: Technician's Pocket Guide

8045, 2/20/03; http://www.hoshizakiamerica.com/TechSupport/TechSpecs/guide%20proof/HOSHI-80045.pdf (accessed May 20, 2014).

4. Cole-Palmer, Thermo Scientific Nalgene Labtop Cooler for 12 (16–17 mm dia.) tubes; http://www.coleparmer.com/Product/Thermo_Scientific_Nalgene_Labtop_cooler_for_12_16_to_17_mm_dia_tubes/EW-44001-30 (accessed May 24, 2014).

5. From advertisement by Thomas Scientific for "Chill bucket with lid"; http://www.thomassci.com/Supplies/Containers/_/1F237F35-9615-45D2-A922-8A556CA69478?q=Ice-Bucket and advertisement by VWR for Lab Armor (R) Chill Buckets, produced by Sheldon; https://us.vwr.com/store/catalog/product.jsp?catalog_number=89409-248 (accessed May 26, 2014).

6. Fisher Scientific, Fisherbrand Polyurethane Ice buckets; http://www.fishersci.com/ecomm/servlet/fsproductdetail_10652_777555__-1_0 (accessed May 24, 2014).

References

Edgerton, David. 2006. *The Shock of the Old: Technology and Global History Since 1900.* Oxford: Oxford University Press.

Homburg, Cornelia, ed. 1996. *Art and Science: Investigating Matter.* St. Louis, MO: Washington Museum Gallery of Art.

Landecker, Hannah. 2007. *Culturing Life: How Cells Became Technologies.* Cambridge, MA: Harvard University Press.

Radin, Joanna. 2013. Latent life: Concepts and practices of tissue preservation in the International Biological Program. *Social Studies of Science* 43:483–508.

5 Nature and the Refrigerating Machine: The Politics and Production of Cold in the Nineteenth Century

Rebecca J. H. Woods

An interest in and an affinity for things cold is as old as humankind itself, at least according to nineteenth-century opinion makers like those writing for *Chambers's Edinburgh Journal*, a widely read general interest magazine published in the middle of the century. "Since the days of that dissipated heathen who, in order to cool the air during an oppressive summer caused mountains of snow to be piled up, and suffered them to melt away, down to the present era, in which there prevails a rage for the thing," one contributor wrote in 1847, "mankind has been incessantly in quest of refrigeratives." The "Ancients," for instance, knew how to harness the cooling effect of the evaporative powers of the sun using porous earthenware jugs: the "invention" of "water-vases formed for that purpose solely" in ancient Egypt had "come down" to contemporary Britain "unaltered in principle" and "with increasing usefulness." Indeed, the "invigorating effect of ice or snow during summer heats" had been recognized "from very early times" (Anon. 1847, 52). Such a quest was driven, according to Victorian observers, by both the practical necessity of food preservation and natural curiosity. For A. J. Wallis-Tayler, the author of a comprehensive and popular "practical treatise on the art and science of refrigeration" that was coming into its own in the late nineteenth century, interest in the effects of cold, and knowledge of "the conservative action of cold upon organic substances," was "probably as old as the existence of human beings" (Wallis-Tayler 1917, 270). Cold, it seemed, was both necessary to human civilization and a universal object of curiosity.

Thus, when engines capable of producing ice and "currents of pure dry cold air" were made operative and increasingly applied to food storage and transport in the 1870s and 1880s, they were heralded as the practical instantiation of this knowledge and the natural conclusion to

humankind's age-old curiosity (Anon. 1878, 6). According to contemporary enthusiasts, the effects of this technology would be manifold. Britons
would be liberated from the vagaries of the seasons and the limits of geography: artificial refrigeration promised ice in the summer and fresh foods
in the winter. On the grandest scale, it promised to bridge the hemispheres
and "arrest the finger of time"; by forestalling putrefaction in meat, refrigeration technology would enable the bodies of sheep and cattle reared as
far away as the Australasian colonies to feed the burgeoning middle and
industrial classes of Great Britain (Anon. 1924, 30). According to Thomas
Mort, a bold standout "amongst all the pioneers in Australia and elsewhere
whose efforts laid the foundations of the frozen meat trade" and one of
the most vocal colonial boosters of the new technology, "Science [had]
drawn aside the veil" when it came to refrigeration (quoted in Critchell
and Raymond 1912). "Climate, seasons, plenty, scarcity, distance, will all
shake hands," Mort predicted, "and out of the commingling will come
enough for all" (20).[1]

Such triumphant positioning of the novel refrigerating engines sounded
a common theme for the time, but it masks a more contingent story of the
production of cold in the nineteenth century. Refrigerating engines represented more than simply the latest—let alone the last—chapter in the story
of a natural human fascination of cold. Rather, they were both the product
of and players in a complex cryopolitical landscape. Artificial cold in part
redrew this landscape, and the idea of a phase change—the physiochemical
process on which mechanical refrigeration relied—illuminates its transformative power. Artificial refrigeration worked by drawing on the abstraction
of heat achieved during a phase change—that is, in the transition from a
liquid to a gaseous state of a substance, or vice versa. In the same way, for
cold itself to become a product meant reconceptualizing it as a *thing* rather
than as an attribute: it became, as the Pure Ice Manufacturing Company
put it in 1878, "a new manufacture" (Anon. 1878). This ontological phase
change depended on contemporary developments in the new science of
thermodynamics, which led to a finer, more precise, and more complete
understanding of heat, and consequently of its inverse: cold.

No less importantly, the cryopolitics of the nineteenth century were
forged within the politics of empire. The ability to control, produce, and
claim ownership over cold was a matter of imperial economics and the politics of colonialism, as well as a challenge to ontologies of matter in its

various states. As Mort's proclamation indicated, artificial refrigeration promised to seamlessly unite the hemispheres—to weave the northern and southern domains of the British Empire into one frictionless whole, transcendent of climatological, cultural, or even seasonal differences. Together, thermodynamism and the imperialism of artificial cold culminated in a significant cryopolitical shift: a new confidence that men—British men, in particular—had triumphed over nature. In producing and disseminating artificial cold, Rule Britannia held sway not only over the four corners of the globe, but over the seasons as well.

A Species of Steam Engine

The machine responsible for the production of cold was the steam engine—that great hallmark of Victorian industry and innovation. This technology worked according to the first law of thermodynamics—the equivalency of work and heat, or that "mechanical energy" was "convertible into heat, and vice versa" (Wallis-Tayler 1917, 8). To this principle, refrigeration technology coupled the second law of thermodynamics, which suggested "that an external agent is necessary to enable heat to pass from a cold to a heated body," as Wallis-Tayler put it (8). This agent could, and did, vary. Nineteenth-century refrigerators operating upon these principles used ammonia, ether, or even simply air. Such machinery harnessed the change in temperature resulting from the compression and subsequent expansion of a gas achieved by the pressure of coal-powered pistons in order to abstract heat from a given agent. That agent could then be channeled into a confined and insulated space to artificially lower its temperature. With "each stroke of the piston of the expansion cylinder" of a cold-air machine, "a jet of intensely cold air is delivered into the chamber," thereby keeping it and its contents at the desired level of artificial refrigeration (Anon. 1893, 424). These machines were thus "in reality" no more than a species of steam engine "with some peculiarities in [their] structure and additional arrangements" (Anon. 1855, 229).[2]

In keeping with the technological enthusiasm that marked the era, the design of equipment for the production of cold was seen as a triumph of engineering. Mechanical refrigerators were the practical instantiations of a principle "discovered in its theoretical form some 150 years ago," according to one unnamed popularizer writing in *Chambers's Edinburgh Journal*: that air or any other gas "which has been compressed, dried, and cooled, will, when suddenly allowed to expand, fall in temperature to a degree below freezing point" (Anon. 1891, 140). The challenge had been "practically to apply this philosophical truth" (140). As historians of science have shown, this "philosophical truth"—even the idea of such a thing as a "freezing point"—was hard won. Furthermore, its discovery was not prior to its practical application (Chang and Yi 2005; Papanelopoulou 2006). Instead, the industrial context was crucial to the development of theoretical knowledge of heat, work, and energy.[3] Steam engines helped to motivate the search for philosophical principles capable of accurately describing their workings. While some interpreted mechanical refrigerators as proofs of a certain philosophy of nature, in fact their creation and use were constitutive of those very theories. In a way, then, the phenomena of steam engines produced their own explanations. This is of consequence because to "manufacture cold," as the Pure Ice Manufacturing Company claimed to be able to do, required the particular understanding of heat born of this scientific-industrial context.

It was here, on heat, that scientific focus remained fixed throughout the nineteenth century. "All we see, all we are, and all the changes that have taken place in our world," proclaimed *Chambers's Journal* in 1858, "seem to be referible [*sic*] to the fact of *heat*" (Anon. 1858, 5). Great attention was focused on understanding and identifying precisely what "heat" described.[4] Common sense suggested that both heat and cold were basic, self-evident facts of existence. Either could be felt simply by a dipping a hand into a bath, or seen registered, by way of its effect on mercury, on a thermometer (Chang 2001, 2004). Experienced in this way, though, heat (or its absence, cold) could only be traced through its instantiation in other objects. Neither heat nor cold as such could itself be isolated.

Over the course of the century, though, as heat came to be defined with increasing precision, a more precise understanding of cold emerged as its inverse. With growing attention to theories of thermodynamism, heat was increasingly understood as "a mode of motion," and cold as the slowing or cessation of that movement (Wallis-Tayler 1917, 8; Chang and Yi 2005; Papanelopoulou 2006). If heat was "a form of molecular energy," then "that negative condition known as cold" was its lack (Wallis-Tayler 1917, 8, 16). In the piston of a steam engine, for example, "each atom is violently agitated, and seeks room for its motion," one commentator explained. The force that drives the piston is thus "nothing more than the combined action of myriads of ultimate atoms seeking their escape." This, in essence, was the work that drove the compression of the refrigerative gases in a "cold-air machine." Pressure from the piston induced a phase change from liquid to gas, which when subsequently allowed to expand within a confined and insulated chamber, absorbed atmospheric heat, thereby lowering the temperature of the chamber. The ultimate effect on whatever was in the chamber—be it meat or milk, fish or fruit—was to produce a state in which "its molecules or atoms are in comparative, if not perfect quietude" (Anon. 1868c, 761).

Even as this process became both better understood and put into practice more and more widely, the essential conceptual components of its workings—heat, cold, temperature—remained elusive.[5] They continued to be palpable, observable, but only through their effects. For cold, such indeterminacy was double, both in its transience as a state or attribute and as something defined primarily in opposition to another concept (heat). In particular, the fact that cold was understood as a lack of heat—as an *absence* of molecular motion—meant that to produce it artificially was to "manufacture" a negative state. Thus the claims of the Pure Ice Manufacturing Company were arguably stranger and even more novel than they appeared.

Artificial Arctics

For nonspecialist audiences, artificial refrigeration was both exciting and potentially unsettling. In the 1870s and 1880s, refrigeration technology and cold stores were a continual topic of general readership magazines, newspapers, trade journals, and specialist texts. Descriptions of cold stores and warehouses, of dockside freezing works and the machines themselves, abounded. Accounts of cold stores, in particular, emphasized refrigeration's power to turn the known order of things on its head. The engines that cooled such spaces were capable of putting extreme artificial cold in stark juxtaposition to naturally occurring atmospheric warmth. "It would ... be difficult," according to *Chambers's Journal*, "to find anything in the way of a contrast more curious than that experienced by a parched and panting Londoner when he steps out of the heat and glare of a broiling day in July or August into an establishment in which one of the latest ice-making machines is at work" (Anon. 1880, 439). Upon leaving "a warm and sunny afternoon to plunge into these Arctic caverns," wrote a contributor to the *Daily News* of the experience, "one is surprised to see one's breath issue forth in a long white stream as on a very raw and misty morning" (Anon. 1882c, 13). Perhaps the only "contrast more curious" than this was found in colonial Australia, where the contrast with the hot exterior environment was most extreme. A visitor to a freezing works in Queensland found "himself at once translated into the atmosphere of a Canadian winter." Here, "a cold, cutting wind is felt, such as is a decided novelty in a climate like ours" (Anon. 1881a, 9).[6]

The disorienting effects of artificial refrigeration were reflected in the architecture of cold storage warehouses. These buildings—"among the most wondrous of recent developments in the river-side enterprises of London," according to a writer for *Chambers's Journal*—were literally "topsy-turvy" (Anon. 1894, 247; Lillingston 1898, 241). A visitor who wished for "the personal experience of the freezing character of the atmosphere" of a cold store entered near the roof, and descended through carefully controlled chambers that decreased in temperature with proximity to the ground level in order to prevent the "irruption of warm outer air into the cold storage chambers" as produce—frozen meat, cheese, and butter—was loaded and unloaded (Anon. 1882a, 12; Wallis-Tayler 1917, 285). In the confined space of a cold store, "winter reign[ed] supreme ... however hot it may be outside" (Anon. 1880, 439). Here, "everything is frozen" (Anon. 1882c, 13). The "innumerable carcases, enveloped each in a calico cover," are covered with a thick hoar frost," and the "ice-covered walls and ceilings and floors with their millions of crystals glistening in the electric light" presented an appearance "picturesque in the extreme" (Anon. 1882c, 13; Anon 1899, 23; Anon. 1881b, 7).

What astonished the journalists who gained access to such enchanted spaces was the capacity of the novel refrigerating engines to replicate the most extreme of natural conditions. Such effects were initially most easily understood in naturalistic terms. Artificial cold was, simply put, just that— a natural effect reproduced by artifice. It begged obvious comparisons to the naturally cold parts of the world, most commonly the Arctic: cold stores were like "Jack Frost's own larder, with the cold of arctic winters in [their] air" (Anon. 1868b, 321). A visitor to such a place "does not feel inclined to stay long in this apartment," according to one observer writing

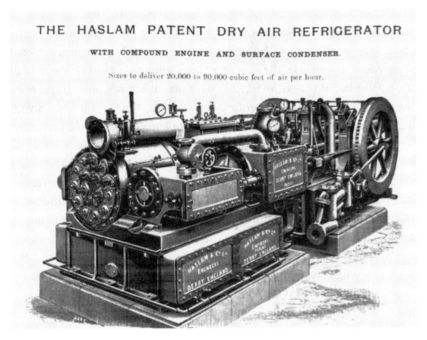

THE HASLAM PATENT DRY AIR REFRIGERATOR

WITH COMPOUND ENGINE AND SURFACE CONDENSER.

Sizes to deliver 20,000 to 90,000 cubic feet of air per hour.

Figure 5.1
A Haslam patent dry air refrigerator. From the *Illustrated Catalogue of Ice Making and Refrigerating Machinery, Manufactured by the Haslam Foundry & Engineering Co., Limited, Incorporated with Pontifex & Wood, Union Foundry, Derby* (1893). Derbyshire Record Office, Haslam Papers, D1333 z/z6.

for the *Otago Times*, one of New Zealand's foremost regional newspapers, "for the cold is so intense that after a very few minutes he begins to fancy he knows something of the severity of an Arctic winter" (Anon. 1882b, 5). These artificial arctics, though, were limited in scale and scope, tightly confined to the holds of the ships that spirited antipodean meat across the oceans, or to the carefully constructed warehouses that stored it upon arrival in Britain, and depended upon effective insulation for their continued existence.[7] Early commentators were aware of these limits: the possibilities of producing artificial climates were recognizably contained in scale and scope. What "nature effects on a large scale," suggested *Chambers's Journal* in 1852, "may reasonably be imitated by man on a more limited one" (Anon. 1852, 257).

Yet such was the power of the new technology, and the enthusiasm it engendered, that this rhetorical relationship, like the physical space of a cold store itself, was very quickly inverted. Where the freezing works of colonial Australasia initially called to mind the cold of the Arctic, Arctic conditions soon came to be understood to resemble a refrigerated cold store. Most striking was the case of a "huge animal ... found imbedded in the ice in Siberia" (Anon. 1852, 257). This mammoth was unearthed before the close of the eighteenth century and in 1852, the fact that "dogs willingly ate of the still existing flesh ... so completely had the cold prevented putrefaction" astonished writers for popular magazines like *Chambers's Edinburgh Journal* and their audiences (257). That deep cold could arrest decay almost indefinitely—that the great beast's flesh remained in "excellent condition after preservation for who knows how many centuries"— surprised Britons in the mid-nineteenth century (Anon. 1881a, 9; Critchell and Raymond 1912, 277). By 1882, however, the same story of the mammoth instead showed "Nature" doing in these icy "parts of the world the work of a refrigerating machine" (Anon. 1882d, 439). Such was the power of artificial refrigeration to produce "ice-house[s] of the chemist's fashioning, completely under man's control," that nature had come to resemble artifice within only a few short years of the development of refrigerating engines (Anon. 1868b, 321).

A Fleeting Commodity

This rhetorical shift gestures to the way in which the ability to recreate the extremes of natural conditions through artificial refrigeration seemed to observers to offer an unprecedented degree of human control over the natural world, even if the actual physical extent of the spaces containing artificial cold in the 1870s and 1880s remained limited. Indeed, artificial cold could even exceed the power of nature at times: in describing the workings of a refrigerating engine, *Chambers's Journal* emphasized how the cooling agent, "having pass[ed] through a spiral pipe in the 'cooler,' ... issues in a frigid torrent which freezes water more rapidly than the severest winter's night" (Anon. 1880, 440). The ability to thus produce and control low temperatures was a liberation from the tyranny of nature, especially in its most immediately relevant incarnation: the North Atlantic ice trade.

Prior to the development of viable refrigerating engines, Britons had relied on natural ice to achieve "the conservative action of cold upon organic substances" (Wallis-Tayler 1917, 270). Local harvesting from rivers and lakes had long been practiced in Britain, weather permitting, but during the first half of the nineteenth century, industrial demand, especially for the fisheries, far exceeded local production (Anon. 1864, 99). Consequently, a vast international trade in ice developed, with New England and Norway supplying the bulk of Great Britain's refrigerative needs. But trade in "so fleeting a commodity" was frustrating: the act of shipping and handling the delicate product necessarily incurred heavy losses from melting despite, in the words of one historian of the trade, "the strenuous attempts to provide effective insulation" (Anon. 1836, 11; David 1995, 53). Indeed, the loss of cargo to melting during a voyage could be considerable, and the longer the voyage, the greater the loss.[8] Consequently, the cost of shipping and subsequent retail prices for natural ice imported to Britain remained high throughout the nineteenth century, independent of fluctuations in supply or demand.[9]

And fluctuate the trade did. Reliance on natural ice had left Britons "quite at the mercy of the weather as concerns a supply" (Anon. 1864, 99). Depending on "naturally-formed ice, gathered only when Jack Frost chooses to make it for us" for food storage, fish mongering, and luxuries such as confectionaries and iced beverages, was "obviously a difficulty to be contended against in mild winters," in the words of one journalist (99).[10] And such variability in supply, according to the Pure Ice Manufacturing Company, meant that "the remedy was worse than the disease, owing to the cost and practical difficulties in the way of constant supplies and storage" (Anon. 1864, 99). The work of refrigerating engines—equally capable of "producing the degree of cold requisite" either for ice manufacture or refrigeration—promised to dissolve these difficulties (Anon. 1878, 21, 6). Creating "a frosty temperature" on command would enable supply to be tailored to demand, and once "cold [could be] easily, cheaply and surely manufactured," promoters promised, Britons would find themselves free from the whims of nature (Anon. 1880, 439; Anon. 1878, 21).[11]

Imperial Appetites

Artificial cold "produced at pleasure" promised to equalize supply and demand for more than just the natural ice trade (Anon. 1858, 5). According to an article in *Chambers's Journal of Popular Literature, Science and Arts*, the ability of cold to·"retard or prevent" the normally "speedy decay of dead

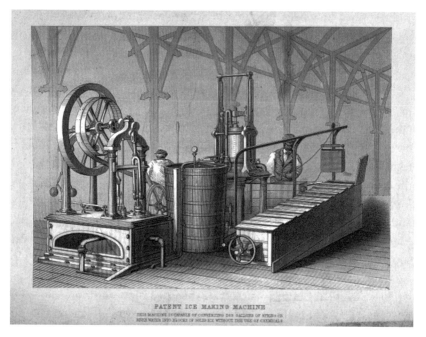

PATENT ICE MAKING MACHINE

Figure 5.2
Two men working at a machine for making ice. Courtesy of the Wellcome Library, London. Library ref.: ICV No. 40432.

animal substances" had great potential to coordinate supply and demand for other perishable articles of food, notably meat (Anon. 1858, 5). Gone were the days "since our country was self-sustaining, its inhabitants content with their own corn and meat," wrote the Pure Ice Manufacturing Company. Nineteenth-century Britons "ransack[ed] the whole globe for clothing and food" and "no country [was] so remote but it minister[ed] in some way to [their] comfort." In attempting to slake such an insatiable thirst for natural resources and raw materials, Britons were "driven to call to [their] aid every resource of science which man's ingenuity or research [could] discover" (Anon. 1878, 3). Foremost among these aids ranked artificial refrigeration, which offered, according to the company, "the only chance of preventing meat from reaching famine prices in our large cities" (6). Indeed, Britain's meat trade was the first practical application of refrigerated storage and shipping, and in many ways motivated the developments in engineering that led to viable refrigerating engines.[12] In this sense, nineteenth-century efforts to produce and manufacture cold were largely, although not exclusively, products of the British Empire. Though French and American innovators were active in the development of the technology, especially early on in the 1850s and 1860s, Britons saw their empire as "the cradle of refrigeration" (Anon. 1924, 32; Arthur 2006, 67–72). It was no accident that the empire also fostered, by the close of the century, "the greatest frozen meat market in the world": its status as such helped stimulate its own "quest of refrigeratives," and both of these characteristics depended on the imperial context (Anon. 1924, 32; Anon. 1847, 52).

One of the most remarkable and intractable problems of empire in the mid-nineteenth century was the question of how to safely and profitably transmit fresh meat from the Australasian colonies to Great Britain. Attempts to successfully transport meat to Britain, such as those spearheaded by James Harrison, a Scot who spent most of his working life in Australia and who became "a pioneer in all kinds of refrigeration," were motivated by the sense that Britain faced a dire shortage of animal protein—a "meat famine" matched in scale only by the excess of sheep and lambs to be found in the Australasian colonies (Bruce-Wallace n.d.; Collins 2000). Indeed, at the same time as Britain's growing and increasingly industrial population outstripped domestic and more proximate (i.e., European) sources of "animal food," the Australasian colonies "were being over-run with cattle and sheep," in the words of R. Ramsay, an early authority on the frozen meat trade (Ramsay 1924, 1721). The only problem was how to get the bounty of the Southern Hemisphere to the waiting, hungry consumers in the metropole—a challenge that was, by the 1870s, "becoming more acute every year" (Ramsay 1924, 1720). For "a meat-consuming people" like the British, the lack of affordable flesh in quantity was indeed painful (Anon. 1868a, 182). A number of schemes—tinning, "boiling down," chemical preservation—had been tried over the years with only limited success. A writer for *Chambers's Journal* reported that between 1691 and 1855 "no fewer than one-hundred-and-ten processes were patented in this country, having for their object the preservation of food from putrefaction" (Anon. 1868c, 760).[13] This observer noted in particular a scheme (popular in the 1860s) to expose the "freshly killed carcass to the action of carbonic acid gas in a closed chamber," which in theory would prevent putrefaction, but which required too many hours and was "attended with more expense and difficulty than should exist in any method which is to be carried out on a large commercial scale" (760).

Such early schemes to preserve meat were unsatisfactory for several reasons. The mere idea of tinned meat or of hunks of mutton and lamb suspended in vats of tallow met with skepticism. In practice, such offerings incited revulsion in British consumers, not least because more often than not, such preservation techniques failed and cargoes spoiled, compounding Britons' disdain for colonial mutton.[14] The arresting effects of cold held out more promise: the logic of imperial trade "must bring … from the Antipodes the superabundant mutton, that in … these wealthy regions find no purchasers," proclaimed a writer for *All the Year Round* in 1868, and ice would supply "the means for effecting this much-desired result." There was reason, this author continued, "for the hopes as well as for the belief that, ere many years have passed, enterprising merchants will be encouraged to convey to our shores fleet loads of the beef and mutton, packed in ice, or chemically frozen, that our population would so gladly purchase" (Anon. 1868a, 182).

Ice itself, though, was too "fleeting a commodity" to preserve flesh for the duration of a lengthy sea voyage (Anon. 1836, 11). Initially, it sufficed to chill beef for the much shorter North Atlantic trade between North America and Great Britain, although ice was replaced by steam-powered refrigerators by the 1870s. To institute the same kind of trade between Australia and Great Britain was more complicated. Early assays from Australia—mostly spearheaded by Thomas Mort—had yet to make the much longer oceanic voyage without spoliation.[15] Such shipments were particularly vulnerable to mechanical failure. For example, in 1873 the *SS Norfolk* carried a cargo of frozen meat in insulated tanks from Melbourne to London. High hopes rode on this "experiment," but owing to faulty construction, the tanks leaked, and "it was ultimately necessary to throw the whole quantity overboard" (Anon. 1873, 93). Even with the aid of refrigerating engines, temporary stoppage during the course of a voyage could damage the vulnerable cargo and ruin profits. With time, adjustments to the technology resulted in more reliable engines, which ameliorated some of these difficulties, including, for instance, the development of dry-air refrigerating engines, which came in a compact shipboard size and operated on air rather than the dangerously flammable ether or corrosive ammonia.[16] Artificial cold—not ice—ultimately solved the British Empire's hemispheric disjuncture in meat supply and demand.

Figure 5.3
Detail of a model refrigeration ship displaying engines and insulated holds. London Museum of Science. Photo by the author, 2011.

Figures 5.4
Detail of a model refrigeration ship displaying engines and insulated holds. London Museum of Science. Photo by the author, 2011.

Even as the engines for freezing meat improved and the growth of the frozen meat trade increasingly overcame the limits of geography, for consumers in metropolitan Britain the implications of the new technology could be hard to swallow. Journalists may have celebrated the ability to contain the Arctic in a cold store, but when it came to the actual consumption of articles of food—especially meat—preserved by the wonders of refrigeration technology, the British public was wary. Frozen meat at first met "with suspicion. … People could not be persuaded that it was, to say the least, palatable," according to E. H. Jackson, writing for *Chambers's Journal* in 1898 (Jackson 1898, 636). If there was one culinary artifact the British prided themselves on, it was meat, and any attempt to alter its traditional preparation and production was likely to agitate public opinion.

By the 1880s, when frozen foreign meat began entering British markets in significant volume, Britain had a well-established, centuries-old commodity chain for meat-confined Northern Europe. This system had undergone much alteration, especially during and after the age of agricultural improvement, but its essence remained largely unchanged. Cattle and sheep were reared at a distance from London (the primary market in Great Britain for meat, as for most other things), purchased by graziers in Buckinghamshire and other adjacent counties, and there "finished" in proximity for the London markets. These proximate sources were supplemented by Scottish and Irish supplies, and sometimes by more distantly raised European animals (Trow-Smith 1957). Overall, this was a highly stratified and a highly specialized system, not to mention one marked by seasonal rhythms: lamb was available at Easter, for example, while older animals—ovine and bovine alike—were available throughout the year according to fluctuating supply.

This was how Britons were used to getting their meat, and the frozen and refrigerated trade presented a serious conceptual (as well as economic) disruption to it, on both temporal and geographic fronts.[17] In the first place, it was by no means self-evident that animals that had "been dead from six to nine months, or even longer," were good to eat (Lillingston 1898, 238). Britons had to be convinced of this, and early popularizers laid stress on the ways in which "dead poultry, and other articles of animal food" were kept "fresh throughout the winter in many rigorous climates" (Anon. 1852, 257). In Russia, the Pure Ice Manufacturing Company assured its readers, "a dead bullock may be seen standing erect, a frozen statue, only to be dismembered with axes and saws: bear's hams are sent frozen across the American continent, and in fact animal tissues can be kept practically any length of time in a frozen condition" (260). James Harrison, for instance, won acclaim at the Melbourne Exhibit of 1873 for demonstrating that "meat kept frozen for months remained perfectly edible" (Bruce-Wallace n.d.). In an economic and culinary context that was based on a relatively quick transition from the paddock to the plate, the preservation of *fresh* (as opposed to salted) flesh without putrefaction required rethinking (Freidberg 2009).

Moreover, the fact that the first applications of artificial cold were to foreign and colonial meat—animals raised and slaughtered abroad, and transported in "what one may call a state of suspended animation"—complicated the already fraught temporal politics of frozen meat. Imports to Britain were composed of beasts that, in the case of Australasia especially, had "cropped pasture land 13,000 miles away" (Anon. 1894, 246; Lillingston 1898, 238). Not only did this pose a problem for the largely chauvinistic British livestock industry, where place and breed were believed to be essential elements of quality, but geography and temporality united to pose a challenge to the familiar rhythm of the seasons that had governed supply (Woods 2013). In the early days of refrigerated trade, it behooved the purveyors of frozen and refrigerated foods to downplay the temporal side-effects of the technology. It was "not generally known, perhaps, that fresh lamb [could] now be obtained out of the refrigeration stores all the year," wrote a contributor to *Chambers's Journal* in 1884, not least because at this point, "the traditional respect for 'seasons' [was] still preserved" by storing up such products until it was seasonally appropriate to make them available (Anon. 1894, 248). This sort of "traditional respect," however, soon eroded under the onslaught of imported colonial meat, and the continued celebration of artificial ice and cold. A mere four years later, according to E. H. Jackson, "what our ancestors deemed impossibilities have become to us matters of every-day simplicity, and articles of food which they regarded as luxuries, only to be obtained at certain seasons of the year, are now, owing to these refrigerating processes, almost every-day articles of diet" (Jackson 1898, 635–636).

Artificial cold allowed imperial Britons to indulge "a lofty disregard of [natural] cold, climate, or season"; the fact that the pastoral industry of Australasia obeyed the seasonality of the Southern Hemisphere, not the

From "*Illustrated London News*," November 19th, 1881.

LANDING AUSTRALIAN FROZEN MEAT FROM SYDNEY, IN THE SOUTH WEST INDIA DOCK, LONDON, FROM THE "S.S. CATANIA." PRESERVED IN TRANSIT BY HASLAM'S REFRIGERATOR.

Figure 5.5

Unloading Australian frozen meat in London. From *Illustrated Catalogue of Ice Making and Refrigerating Machinery* (1893). Originally printed in the *Illustrated London News* (November 19, 1881). Derbyshire Record Office, Haslam Papers, D1333 z/z6.

Northern, and thus ewes giving birth in the Austral spring did so only a few months before Great Britain's Christmas market, no doubt helped in over-turning "traditional" seasonality (Anon. 1864, 101).

A Most Beneficial Application

The material effects of artificial refrigeration were felt most immediately and most powerfully in its application to Britain's meat supply. Beyond this realm, though, artificial cold also promised to ameliorate another imperial problem: that of tropical climates and the challenges they posed for British bodies. Thermodynamically, cold could be understood as what Wallis-Tayler had called that mere "negative condition"—as a lack of heat—but the earth was "for the greater part a sunburnt one," and outside of the chilly regions of northern Europe, imperial Britons experienced the distinctly hot tropics as places climatically lacking in cold (Wallis-Tayler

1917, 16; Anon. 1858, 235). The "wide tropical belt" encompassing much of India, the West Indies, and even Australia, and "embracing the greater portion of the earth's peopled surface and the vast majority of its inhabitants," suffered, according to *Chambers's Journal*, "an almost continual oppression and distress from its exposure to the unmitigated glare" of the sun (Anon. 1858, 235). Would-be European colonizers thus experienced conditions most unlike the climate of the British Isles. The logic of imperial acclimatization shifted over time, as Mark Harrison has shown, from optimism about the adaptability of Europeans to tropical environments in the early modern period to racial pessimism in the nineteenth century (Harrison 1999). Despite changing explanatory mechanisms, though, that the heat of the tropics constituted a danger and a shock to European bodies remained universally accepted.

In India, in particular, the differences between northern Europe's cool temperate locales and the extreme heat of more equatorial climes were understood to be extremely hazardous starting in the eighteenth century. For Fleming Martin, for example, the chief engineer in Bengal in the 1760s, the "intense and uncommon heat in this climate [was] ... almost insufferable" (Martin and Martin 1767, 218). Even when the climate did not prove fatal, it "soon exhaust[ed] a person's health and strength, though ever so firm in constitution, as [was] visible in every countenance after being here twelve months" (219). Aided by the seasonality of "pucker fevers," which broke out with the heat of summer, the tropics increasingly came to be defined as inherently pathogenic (Barker and Cavendish 1775, 206; Harrison 1999).[18] Moreover, as racial science evolved in the nineteenth century toward understandings of "type" as relatively fixed rather than the product of environment, acclimatization to local colonial conditions ceased to be the desideratum it had been in an earlier era (Arnold 1996; Anderson 2003, 2006). Increasingly, those who held out against the "malignity of the climate" in tropical colonies risked compromising their racial identity and becoming "a kind of hybrid, inferior to both" their European origins and "native" Indians (Martin and Martin 1767, 218; Harrison 1996, 1999; Adamson 2012). Time spent in the tropics was a thus threat to both the health and the whiteness of European colonizers, and one that only grew over the course of the nineteenth century as European colonialism solidified throughout India, Australia, and much of Africa (Harrison 1996; Anderson 1992, 2006; Livingstone 1999).

Colonists were not entirely helpless in the face of tropical climatic hazards. Efforts to ameliorate local climates, or at the very least to minimize the harmful effects of unfamiliar conditions, had been an integral part of the process of colonization since at least the early modern period (Kupperman 1984; Valencius 2002; Zilberstein 2016). Alterations in dress, for example, or the avoidance of "indulgent" or "intemperate" behavior could mitigate the bodily effects of tropical environments. Local environments could also be modified to enhance their salubrious aspects and mitigate their harmful ones—swamps drained, salubrious species introduced, and supposedly harmful ones contained or eradicated (Harrison 1999, 87–88). In India, as territorial rule intensified over the course of the nineteenth century, environmental improvement and landscape alterations increasingly became important tools in the establishment and perpetuation of colonial rule. Eucalypts—a species native to colonial Australia and believed to be capable of purifying the atmosphere—were introduced to the subtropics in the hopes of improving conditions for Europeans (Beattie 2012). The establishment of elevated hill stations, in particular, became an increasingly important strategy for countering the ill effects of the tropics on European bodies (Harrison 1999, 124–132; Kennedy 1996).[19] "Hill sanataria" offered "a means of not only rearing children, but of preserving the health of adults," according to "an Anglo-Indian" writing for *Chambers's Journal* (Anglo-Indian 1875, 373). The "diverse residences for Europeans" scattered throughout India's mountainous regions were "situated amidst the most magnificent scenery," and allowed, "according to altitude, any one [to] choose a climate to his taste" (373). If colonists lacked periodic recourse to hill stations or sojourns at "Home" undertaken for their health, contemporary experts believed that continued British rule in India would be impossible (Harrison 1999, 139–141).

Retreat to hill stations, though, offered only temporary respite from the extremity of India's climate, and the effects of landscape alterations or changes to personal comportment were an incomplete solution. Throughout the nineteenth century, little could be done to alter what was perhaps the most fundamental challenge of tropical environments: the heat. The persistent ill effects of tropical conditions in India, as in Australia and elsewhere, constituted a significant obstacle to effective colonial rule.[20] Expressing a sentiment common to contemporary observers, for instance, *Chambers's Journal* reported in 1858 that "we ourselves know a lady whose husband was forced to resign a valuable governorship in a tropical climate owing to her health giving way" (Anon. 1858, 4). Refrigeration technology proposed to change all of this more effectively than landscape alteration and more persistently than high-altitude retreats. Just as it promised to equalize supply and demand between the various parts of the British Empire, the technology of artificial cold and manufactured ice offered to solve the problem of tropical climates. Standardizing the temperature of the Empire was one of the most exciting potential applications of the novel technology, even though such climatic applications were far more specula-tive than their use in the frozen meat trade. In the place of temporary and necessarily recursive recourse to hill stations, artificial cold promised what would ultimately be its "most beneficial application," early commentators believed: the "cooling of apartments ... in tropical regions" (Anon. 1858, 6). Instead of tatties, punkahs, and thermantidotes—various contrivances developed for regulating interior temperatures in parts of India during the hot season—with an unlimited supply of artificial cold, "rooms can be cooled by placing vessels of ice in them" (Anon. 1864, 101).[21]

Importantly, natural ice was not enough to provide the requisite cold and ice to effectively cool the climate of India's confined habitations: only artificial refrigeration could achieve this. While ice could be produced under the right conditions in the mountainous "upper provinces," Britons believed indigenous cooling systems were inadequate and insufficiently sophisticated to solve the problem of the "intense and uncommon heat in this climate" (Anon. 1836, 10; Martin and Martin 1767, 218).[22] "The effect which cheap supplies of ice may possibly have in enabling the white man to live and work in tropical climates," on the other hand, held out more possibility: "If beverages can be cooled by means of ice; if meat and other articles of food can be preserved in good condition for some time by its agency ... if these things be so, then some, at least, of the miseries that press upon the white man in a hot climate might be alleviated, and we might then really see what northern muscles can effect in southern regions" (Anon. 1864, 101).

For enthusiasts of artificial refrigeration, a reliable supply of cold and ice was all that was needed to make India and other tropical colonies habitable and risk-free for British colonists. It would not be long, the *Illus-trated London News* predicted, before "ice, within the tropics, [would] ... be looked upon as a necessary of life" (quoted in Anon. 1858, 6). Here, in "hot climates," was where "the full value of the invention of artificial cold would be felt," a writer for *Chambers's Journal* predicted (6). The unfortunate governor's wife, *Chambers's Journal* proclaimed, needed "nothing more than a sufficient supply of ice ... to enable her to remain" in the colonies (4).

Conclusion

In the 1880s, Britain's trade in frozen meat, though fast-growing, was in its infancy, and the climatic applications of artificial cold, despite the enthusiastic rhetoric of its promoters, outside of cold stores and refrigerated ships, was largely speculative. A generation later, in June 1924, when the International Institute of Refrigeration convened the Fourth International Congress of Refrigeration, artificial refrigeration in its industrial guise was roughly forty years old, the situation was much changed, and refrigeration, it seemed, had borne out its early promise. Over the course of the five-day meeting, ceremonial banquets—the menus of which were reported in the *Proceedings of the Fourth International Congress of Refrigeration* to have had "a great debt to pay to cold and its various applications"—offered repeated opportunity for representatives from participating nations to reflect on what they saw as the profoundly unifying geopolitical effects of refrigerated trade (Anon. 1924, 38). A spirit of international cooperation permeated the Congress: "In this parliament of Refrigeration," declared Gordon H. Campbell, "the Old and the New Worlds are united" (32). By this point, it seemed that nothing lay beyond the improving reach of artificial cold. Refrigeration could overcome geography, conquer time, even save humanity from itself. World War I was a recent and vivid reminder of "what peril menaced humanity," the French delegate to the International Congress reflected, and had demonstrated "the need for using refrigeration to limit the extent of this peril, and confine its reach" (38). Proponents of the trade gathered at this Congress argued that the technology and its commercial uses had lived up to its potential to redefine geography and temporality. In bringing the produce of the Southern Hemisphere to Europe and the United Kingdom, it was "not too much to say," according to Sidney J. Webb, president of Great Britain's Board of Trade, "that it is through the discovery of the possibility of refrigeration that we owe, if not the present density of population in Europe, at any rate its power to exist at a high standard of subsistence" (30).

Meanwhile, on the other side of the world, refrigeration technology was credited with rescuing the agropastoral industries of colonial Australasia. For the first time, "the application of cold air to foodstuffs" allowed colonial producers to profit from the meat as well as the wool of their vast flocks of sheep, and according to Sir Thomas Mackenzie, New Zealand's representative to the Congress, it "[had] meant the very breath of life" to the colonial economies by stimulating the colony's export industry when global wool trade flagged (46). Even more significantly, the preservative effect of artificial cold on perishable victuals had apparent and profound temporal implications. With the power of refrigeration to "suspend animation," to slow or even halt decay, the men of science behind its technological development and the men of commerce responsible for its widespread practical application could claim, Webb generously proclaimed, "to have arrested the finger of time" (30). Mort's prediction, it seems, had come to pass: thanks to refrigeration, "climate, seasons, plenty, scarcity, distance," had "all shake[n] hands, and out of the commingling [had] come enough for all" (quoted in Critchell and Raymond 1912, 20). "Science" had allowed imperial Britons to transcend natural limits, and had, according to an article entitled "Ice-Making" published in *Chambers's Journal of Popular Literature, Science and Arts*, "succeeded in producing and maintaining, in contempt of almanacs, and in defiance of dog-stars and noonday suns … wintry temperatures" (Anon. 1880, 439).

Yet the cryopolitical developments of the nineteenth century were not an unmitigated good. In the longer term, the promise of artificial cold—to unify the globe, to erase the seasons, to bring the tropics to heel—has not been borne out. Cold never smoothed the path for imperial rule in India: that it could, or should, ever do so was a fiction. Attempts to smooth away temporal and geographical disjunction in the nineteenth century produced new forms of colonial or quasi-colonial dependency and exploitation in the twentieth, as Susanne Freidberg has shown (Freidberg 2010). Subsequent essays in this volume demonstrate the way in which the preservative promise of refrigeration technology is and has been continually deferred, almost regardless of context or application. The roots of our current cryopolitical regimes were put down in the early world of mechanical cold—in the figurative change of state that the concept of cold itself underwent, from attribute to object, in the quest to overcome the limits of climate and the seasons, and in the imperial politics and the politics of meat in which these processes took place. In this, the thermodynamism of the nineteenth century laid the foundation for the elaboration of the cryopolitical in times to come.

Notes

1. Speech given Sept. 2, 1875.

2. Wallis-Tayler (1917, 19) described refrigerating engines as heat pumps operating in reverse.

3. Smith and Wise 1989 remains authoritative. In a distillation of his coauthored monograph, Smith argues that the context of Scottish natural philosophy in the early to mid-nineteenth century in which William Thomson (later Lord Kelvin) operated and the industrial context that inspired James Joule and others were crucial to the development of a distinctly "Northern British" school of thermodynamics that became widely accepted by the close of the century. In adopting this position, Smith and others argue against both the nineteenth-century view that theory was usually prior to practice, and Thomas Kuhn's ahistorical model of "simultaneous discovery" (Smith 2002, esp. p. 290. See also Kuhn 1959; Myers 1985).

4. The Northern British school of thermodynamics relied on the motion theory of heat as opposed to the "action at a distance" school of thought that predominated in France (Smith 2002, 293).

5. As Hasok Chang points out, attempts to define or measure temperature were problematically circular: "How," he asks, "can we test whether the fluid in our thermometer expands regularly with increasing temperature, without a circular reliance on the temperature readings provided by the thermometer itself?" (Chang 2004, 4; see also Chang and Yi 2005, 290).

6. Queensland, the northernmost province in Australia, is also the most tropical part of the continent, lying north of the Tropic of Capricorn.

7. Materials used for insulation varied, from cork to "asbestos, cotton-wool, sheep's wool, pine-wood, loam, gas works breeze, coal ashes, sawdust, hair felt, lamp black, mica, paper, fine cinders, pitch, etc." (Wallis-Tayler 1917, 298; see also Critchell and Raymond 1912, 341).

8. The first shipment of natural ice from New England to India in 1833, for instance, during which the vessel was at sea for four months and seven days, resulted in the loss of 80 tons of ice out of an original 180, or 44 percent of the cargo. However, heavy losses in transit were not a deterrent to the establishment or growth of the trade; rather they were seen as an unavoidable cost of the trade, which necessarily brought ice, the product of cold, northern climes, into warmer realms (Anon. 1836, 11; Anon. 1864, 100; David 1995).

9. David shows conclusively that the high cost of shipment and handling effectively outweighed any impact of fluctuating supply and demand on the retail price of natural ice (David 1995, esp. 53).

10. David argues that the ice trade "was clearly at the mercy of the weather: a mild winter reduced supply, a cool summer reduced consumption" (David 1995, 54).

11. Indeed, David argues of the Anglo-Norwegian ice trade that "the growth of artificial refrigeration ... was the single most important factor in its decline" (David 1995, 52).

12. A great deal of this innovation took place in Australia, although the first effective shipboard refrigerating engines were produced in Scotland by the Bell Coleman Company in 1877 (Arthur 2006, 75).

13. Between 1819 and 1876, 137 patents for mechanical refrigeration alone were registered in Great Britain (Critchell and Raymond 1912, 425–431).

14. A most spectacular instance of this occurred in 1851 when government inspectors found that fewer than 100 out of 6,000 canisters of tinned meat purchased less than a year earlier by the British admiralty remained fit to eat. The contents were found "in a state of such loathsome putridity as to render the office of the examiners a terrible one" (Anon. 1852, 258).

15. For the role of Australian engineers and entrepreneurs in the development of refrigeration technology, see Arthur 2006.

16. Sir Alfred Haslam developed the first shipboard refrigerator to run on air rather than ammonia (Critchell and Raymond 1912, 37, 372–373; Anon. 1891).

17. These rearrangements had important consequences for matters of taste, as well. See Perren 2006; Woods 2012. For the economic effects of the frozen meat trade, see Higgins 2004.

18. "Pucker fever" was the contemporary name for cholera; in colonial West Africa and the Caribbean, yellow fever and malaria constituted the primary feverish threats to European colonizers. For European mortality in the tropics, see Curtin 1989.

19 The significance of upland retreats for European tropical colonialism held in the French Empire as well. See Jennings 2006.

20. Warwick Anderson (2006) analyzes climatic impediments to Anglo sovereignty in Australia in terms of its impact first on white British colonialism and subsequently on Australia's racial nationalism, also dependent on whiteness.

21. Tatties were aromatic woven reed screens hung in doors and windows and sprinkled with water to enable evaporation to cool the interior atmosphere of a house, while punkahs and thermantidotes were large, hand-operated fans, respectively hung from the ceiling and resembling a large bellows. See Anglo-Indian 1881, 686–687.

22. The method of producing ice in the highlands of India relied on shallow clay pans resting in beds of straw, in which water exposed to overnight temperature changes evaporated, forming ice. For early accounts of the process, see Barker 1775, 252–257; Williams 1793a, 56–58, 1793b, 129–131.

References

Adamson, George C. D. 2012. "The langour of the hot weather": Everyday perspectives on weather and climate in colonial Bombay, 1819–1828. *Journal of Historical Geography* 38:143–154.

Anderson, Warwick. 1992. Climates of opinion: Acclimatization in nineteenth-century France and England. *Victorian Studies* 35 (2): 135–157.

Anderson, Warwick. 2003. The natures of culture: Environment and race in the colonial tropics. In *Nature in the Global South: Environmental Projects in South and Southeast Asia*, ed. Anna Lowenhaupt Tsing. Durham, NC: Duke University Press.

Anderson, Warwick. 2006. *The Cultivation of Whiteness: Science, Health, and Racial Destiny in Australia*. Durham, NC: Duke University Press.

Anglo-Indian. 1875. The European in India. *Chambers's Journal of Popular Literature, Science and Arts* (June 12): 371–374.

Anglo-Indian. 1881. India in the Hot Weather. *Chambers's Journal of Popular Literature, Science and Arts* (October 22): 686–688.

Anon. 1836. The ice trade between America and India. *Mechanics' Magazine, Museum, Register, Journal, and Gazette* 25 (April 9–September 24): 10–12.

Anon. 1847. Artificial cold. *Chambers's Edinburgh Journal* (July 24): 52–54.

Anon. 1852. Preserved meats and meat-biscuits. *Chambers's Edinburgh Journal* (October 23): 257–260.

Anon. 1855. American jottings: Late inventions and projects. *Chambers's Journal of Popular Literature, Science and Arts* (April 14): 228–231.

Anon. 1858. Artificial ice-making. *Chambers's Journal of Popular Literature, Science and Arts* (July 3): 4–6.

Anon. 1864. Ice-culture. *Chambers's Journal of Popular Literature, Science and Arts* (February 13): 99–101.

Anon. 1868a. Ice. *All the Year Round* (August 1): 180–182.

Anon. 1868b. Australian mutton. *All the Year Round* (September 12): 319–321.

Anon. 1868c. Our meat-supply. *Chambers's Journal of Popular, Literature, Science and Arts* (November 28): 759–762.

Anon. 1873. Science: Frozen Australian meat. *London Reader* 22 (551): 93.

Anon. 1878. *Cold: A New Manufacture*. London: Field & Tuer.

Anon. 1880. Ice-making. *Chambers's Journal of Popular Literature, Science and Arts* (July 10): 439–441.

Anon. 1881a. Meat-freezing works. *Queenslander* (October 23). Reprinted in *Haslam's Patent Dry Air Refrigerators* (pamphlet): 9. Derbyshire Record Office D1333 z/z 5.

Anon. 1881b. Arrival of frozen meat from Australia. *Daily News* (October 24). Reprinted in *Haslam's Patent Dry Air Refrigerators* (pamphlet): 7. Derbyshire Record Office D1333 z/z 5.

Anon. 1882a. Haslam's patent refrigerator at Woodside Lairage. *Liverpool Daily Courier* (July 5). Reprinted in *Haslam's Patent Dry Air Refrigerators* (pamphlet): 12. Derbyshire Record Office D1333 Z/Z 5.

Anon. 1882b. Haslam's patent dry air refrigerator in New Zealand. *Otago Times* (August 5). Derbyshire Record Office D1333 z/z 3.

Anon. 1882c. A new experiment in food supply. *Daily News* (September 26). Reprinted in *Haslam's Patent Dry Air Refrigerators* (pamphlet): 13. Derbyshire Record Office D1333 Z/Z 5.

Anon. 1882d. Frozen food. *Chambers's Journal of Popular Literature, Science and Arts* (July 14): 437–439.

Anon. 1891. Sir Alfred Haslam, K.T., J.P.: A sketch of his career. In *The Queen's State Visit to Derby 21 May 1891. Illustrated Memorial Volume*. Derby: W. Hobson. Derbyshire Record Office D1333 z/z 6.

Anon. 1893. Notes on current science, invention, and discovery. *Leisure Hour* (April): 421–424.

Anon. 1894. Australian meat-trade. *Chambers's Journal of Popular Literature, Science and Arts* (April 21): 245–248.

Anon. 1899. Our meat-supply. *Chambers's Journal of Popular Literature, Science and Arts* (August 26): 615–616.

Anon. 1924. *Proceedings of the Fourth International Congress of Refrigeration*, June 16–21, 1924, vol. 1. London: International Refrigerating Congress Movement, British Cold Storage and Ice Association.

Arnold, David. 1996. *The Problem of Nature: Environment, Culture and European Expansion*. Oxford: Blackwell.

Arthur, Ian. 2006. Shipboard refrigeration and the beginnings of the frozen meat trade. *Journal of the Royal Australian Historical Society* 92 (1): 63–82.

Barker, Robert, and Henry Cavendish. 1775. An account of some thermometrical observations, made by Sir Raboert Barker, F.R.S. at Allahabad in the East Indies, in Lat. 25° 30′ N. During the year 1767, and also during a voyage from Madras to England, in the year 1774. Extracted from the original journal by the Hon. Henry Cavendish, F. R. S. *Philosophical Transactions* 65:202–206.

Barker, Robert. 1775. The process of making ice in the East Indies. By Sir Robert Barker, F. R. S. in a letter to Dr. Brocklesby. *Philosophical Transactions* 65:252–257.

Beattie, James. 2012. Imperial landscapes of health: Place, plants, and people between India and Australia, 1800s–1900s. *Health and History* 14 (1): 100–120.

Bruce-Wallace, L. G. n.d. Harrison, James (1816–1893). In *Australian Dictionary of Biography*. National Centre of Biography, Australian National University. http://adb.anu.edu.au/biography/harrison-james-2165/text2775 (accessed November 14, 2013).

Chang, Hasok. 2001. Spirit, air, and quicksilver: The search for the "real" scale of temperature. *Historical Studies in the Physical and Biological Sciences* 31 (2): 249–284.

Chang, Hasok. 2004. *Inventing Temperature: Measurement and Scientific Progress.* Oxford: Oxford University Press.

Chang, Hasok, and Sang Wook Yi. 2005. The absolute and its measurement: William Thomson on temperature. *Annals of Science* 62 (3): 281–308.

Collins, E. J. T. 2000. Food supplies and food policy. In *The Agrarian History of England and Wales*, vol. 7, part 1. Ed. E. J. T. Collins, 33–71. Cambridge: Cambridge University Press.

Critchell, James Troubridge, and Joseph Raymond. 1912. *A History of the Frozen Meat Trade: An Account of the Development and Present Day Methods of Preparation, Transport, and Marketing of Frozen and Chilled Meats.* London: Constable & Company.

Curtin, Philip D. 1989. *Death by Migration: Europe's Encounter with the Tropical World in the Nineteenth Century.* Cambridge: Cambridge University Press.

David, Robert. 1995. The demise of the Anglo-Norwegian ice trade. *Business History* 37 (3): 52–69.

Freidberg, Susanne. 2009. *Fresh: A Perishable History.* Cambridge, MA: Belknap Press of Harvard University Press.

Freidberg, Susanne. 2010. Freshness from afar: The colonial roots of contemporary fresh foods. *Food & History* 8 (1): 257–277.

Harrison, Mark. 1996. "The tender frame of man": Disease, climate, and racial difference in India and the West Indies, 1760–1860. *Bulletin of the History of Medicine* 70 (1): 68–93.

Harrison, Mark. 1999. *Climates and Constitutions: Health, Race, Environment and British Imperialism in India, 1600–1850.* New Delhi: Oxford University Press.

Higgins, David M. 2004. "Mutton dressed as lamb?" The misrepresentation of Australian and New Zealand meat in the British market, c. 1890–1914. *Australian Economic History Review* 44 (2): 161–184.

Jackson, E. H. 1898. Cold storage and our food supply. *Chambers's Journal of Popular Literature, Science and Arts* (September 3): 636.

Jennings, Eric T. 2006. *Curing the Colonizers: Hydrotherapy, Climatology, and French Colonial Spas.* Durham, NC: Duke University Press.

Kennedy, Dane. 1996. *The Magic Mountains: Hill Stations and the British Raj.* Berkeley: University of California Press.

Kuhn, T. S. 1959. Energy conservation as an example of simultaneous discovery. In *Critical Problems in the History of Science,* 321–356, ed. M. Clagett. Madison: University of Wisconsin Press.

Kupperman, Karen Ordahl. 1984. Fear of hot climates in the Anglo-American colonial experience. *William and Mary Quarterly* 41 (2): 213–240.

Lillingston, Leonard W. 1898. Frozen food. *Good Words* (January): 237–244.

Livingstone, David N. 1999. Tropical climate and moral hygiene: The anatomy of a Victorian debate. *British Journal for the History of Science* 32 (1): 93–110.

Martin, William, and Fleming Martin. 1767. Letter to Mr. Dacosta, Librarian, &c. to the Royal Society, from Mr. William Martin; Containing an extract of a letter from his son at Bengal, on the heat of the climate. *Philosophical Transactions* 57:217–220.

Myers, Greg. 1985. Nineteenth-century popularizations of thermodynamics and the rhetoric of social prophecy. *Victorian Studies* 29 (1): 35–66.

Papanelopoulou, Faidra. 2006. Gustave-Adolphe Hirn (1815–90): Engineering thermodynamics in mid-nineteenth-century France. *British Journal for the History of Science* 39 (2): 231–254.

Perren, Richard. 2006. *Taste, Trade and Technology: The Development of the International Meat Industry since 1840.* Aldershot, UK: Ashgate.

Ramsay, R. 1924. The rise of the world's refrigerated meat traffic, and its effect on the resources of the various countries of meat supply. In *Proceedings of the Fourth International Congress of Refrigeration,* June 16–21, 1924, vol. 2: 1720–1733. London: International Refrigerating Congress Movement, British Cold Storage and Ice Association.

Smith, Crosbie, and M. Norton Wise. 1989. *Energy and Empire: A Biographical Study of Lord Kelvin.* Cambridge: Cambridge University Press.

Smith, Crosbie. 2002. Force, energy, and thermodynamics. In *The Cambridge History of Science*, vol. 5: *The Modern Physical and Mathematical Sciences*, 289–310. ed. Mary Jo Nye. Cambridge: Cambridge University Press.

Trow-Smith, Robert. 1957. *A History of British Livestock Husbandry*, vol. 1. London: Routledge and Kegan Paul.

Valencius, Conevery. 2002. *The Health of the Country: How American Settlers Understood Themselves and Their Land*. New York: Basic Books.

Wallis-Tayler, A. J. 1917. *Refrigeration, Cold Storage and Ice-Making: A Practical Treatise on the Art and Science of Refrigeration*, 5th ed. London: Crosby, Lockwood and Son.

Williams, John Lloyd. 1793a. Account of the method of making ice at Benares. In a letter to William Marsden, Esq., F. R. S. from John Lloyd Williams, Esq., of Benares. *Philosophical Transactions of the Royal Society of London* 83:56–58.

Williams, John Lloyd. 1793b. Additional observations on the method of making ice at Benares. In a letter to William Marsden, Esq., F. R. S. from John Lloyd Williams, Esq., of Benares. *Philosophical Transactions of the Royal Society of London* 83:129–131.

Woods, Rebecca J. H. 2013. The herds shot round the world: Native breeds and the British Empire, 1800–1900. Unpublished PhD. dissertation, Massachusetts Institute of Technology.

Woods, Rebecca J. H. 2012. Breed, culture, and economy: The New Zealand frozen meat trade, 1880–1914. *Agricultural History Review* 60 (2): 288–308.

Zilberstein, Anya. 2016. *A Temperate Empire: Making Climate Change in Early America*. Oxford: Oxford University Press.

6   Stockpiling as a Technique of Preparedness: Conserving the Past for an Unpredictable Future

Frédéric Keck

Introduction: Frozen Corpses, Chilled Chickens

In 1997, two simultaneous events shook the tiny world of influenza researchers. One set of events unearthed frozen corpses to the surface to explore a century-old virus that had killed millions of people. The second saw the emergence of a new virus whose threat would be mitigated by the use of chilled bodies.

The first event was triggered by the publication in the journal *Science* of the genetic sequence of the hemagglutinin protein of the virus that had caused the 1918 "Spanish flu," estimated to have killed between 20 and 50 million people worldwide (Taubenberger et al. 1997). Jeffery Taubenberger, the first author on the paper, had been working as a molecular biologist at the Armed Forces Institute of Pathology in Washington, DC, where lung specimens from soldiers who died of flu after the war had been preserved. Taubenberger's publication led to an upsurge of new research on this old strain of flu virus. Pathologist Johan Hultin had previously unearthed victims of the 1918 flu buried in the permafrost of Brevig Mission, Alaska, but he failed to find live viruses at that time. Following his advice, Taubenberger returned to Brevig, where he and his team found DNA fragments in the frozen lung of a young obese Inuit woman. From these fragments, researchers were able to amplify and publish the full sequence of the 1918 flu virus.[1] They named the woman "Lucy."

The 1918 flu is often regarded as a "forgotten" pandemic (Crosby 2003). The story of its recovery from "nature's freezer" is told as the beginning of a new era of research, made possible through the use of genetic sequencing of inadvertently preserved tissue to reveal mysteries that had remained hidden for almost a century. With the possibilities of PCR (polymerase chain reaction) and reverse genetics (where gene function is inferred from its sequence), many claimed that obtaining actual fragments of the 1918 virus could give a complete picture of how a seasonal respiratory disease caused one of the deadliest pandemics in history. This unlocking of nature's mysteries by new decoding methods also provoked concern about the ability for genetic data about a "killer virus" to be misused. One of the most disturbing scenarios imagined was that bioterrorists would use the genetic sequence to resurrect the 1918 flu virus.

However, concern about the threat of reanimated viruses waned as scientists became frustrated in their inability to pinpoint a genomic explanation for what had made the 1918 flu virus so lethal. Following Taubenberger's initial publication of its genome, research on the specific gene that researchers hoped would explain the lethality of the 1918 "H1N1" virus, led by Peter Palese at Mount Sinai Hospital in New York, was deemed unsuccessful (Caduff 2012, 2014). The gene thought to contribute to viral pathogenicity, PB1, was identical in the 1918 reconstructed virus and modern flu viruses. This meant that a genetic explanation might not be sufficient to explain why the 1918 strain had been so deadly. While this provided some degree of assurance that it would not be easily weaponized, it also meant that any hopes of using the old virus to produce a novel antiviral medication were stymied. The mysteries of the severity and evolution of pandemic flu viruses remained obscure.

That same year, 1997, a new flu virus called "H5N1"[2] was identified in birds and humans in Hong Kong. Between May and December, it infected eighteen people, six of whom died, and it was estimated to have infected 20 percent of the poultry in the territory. This flu virus jumped directly from birds to humans, without going through the "mixing vessel" of pigs, a species with receptors for both avian and human flu viruses. Scientists speculated that this jump of the species barrier could explain the pathogenicity of the H5N1 virus. They theorized that young immune systems seemed to overreact to a virus deriving from a radically different organism through a phenomenon called "cytokine storm" (Peiris, de Jong, and Guan 2007).[3] A potential solution to the mystery of pandemic flu viruses was thus proposed: they circulate among animals before jumping to humans with catastrophic effects on their immune system, before becoming seasonal and less pathogenic, in a cyclical mode.

World Health Organization (WHO) experts who went to Hong Kong to investigate the case, as well as local scientists such as Kennedy Shortridge, head of the Department of Microbiology of Hong Kong University, thought it might be possible to stop the virus before it became pandemic—that is, before it began spreading from human to human and not only from birds to humans—by producing a vaccine for this emerging avian flu virus (Shortridge and Stuart-Harris 1982). In an ironic twist, this proved technically impossible, as vaccines are cultivated in chicken embryos, which are killed by the avian flu virus. The alternative solution was to eradicate the virus in its "animal reservoir." Consequently, all poultry living in the Hong Kong territory were culled in November 1997. The Hong Kong government (recently handed over to the authority of Beijing) encouraged the consumption of chilled poultry imported from mainland China. Chilled poultry doesn't present risks of transmission since the virus cannot live in a dead carrier. Imported, prekilled poultry allowed food and hygiene administrations to control circulation of meat through the cold chain and, in the process, minimize human–animal contact.[4]

We thus have two parallel sets of 1997 flu-related events: the hope that resurrection of frozen bodies of human victims of a past flu virus could, in turn, resurrect a dormant virus, and the effort to curtail a future pandemic virus by circulating poultry through the cold chain rather than allowing consumers to purchase live birds at the market. These events present inverse relations between cryopolitics and biosecurity: while the potential unlocking of flu virus from frozen corpses long preserved in "nature's freezer" raised concerns about bioterrorism, chilled chicken mitigated the risks of "nature as a bioterrorist threat" through the use of biosecurity practices. These two events also present inverse relations between life and death: while the unearthing of frozen corpses was a messianic form of "making the dead live," the massive culling of poultry was a sacrificial gesture of "making the living die." These apparently opposing approaches to knowing and controlling pandemic flu viruses were undermined by the constant flux of new viruses through migratory wild birds or human visitors coming from China into Hong Kong. Some have argued that avian influenza is a form of "revenge" of animals against the humans who try to control them through livestock production (Greger 2006). This subversive potential of emerging "wild" viruses stored in fridges blurs the opposition between nature and the laboratory as well as between life and death. These two events are alternate sides of the same cryopolitical coin: one side presents a threat toward humans and the other toward animals. I look at the material the coin is made of, to ask questions about its value.

In this essay, I examine the phenomenon of viral mutations in birds and humans through the framework of cryopolitics. Between the "naturally" frozen humans, resurrected to study old flu, and the poultry that has been "artificially" refrigerated to stave off new flu, I am interested in another regime—or form of currency—for managing knowledge about and risk related to flu: the collection and cryopreservation of flu samples in birds and humans, the production of vaccines it entails, and the politics of preparedness it invokes. How does cryopolitics capture the unfolding mutation of flu viruses, and how does it address its scientific and political uncertainties (Radin 2014, 2015)? I want to show that, rather than the religious vocabulary of messianic resurrection or sacrificial expiation, these practices are more appropriately understood through the "magical" vocabulary of hunters. Neither a leap of faith into a cryonic future (Bunning, this vol.) nor the arbitrary excavation of victims from the past, the analysis of jumps between species reveals a present field of mutations connecting humans and animals. I will therefore leave aside frozen corpses and chilled chickens to open the fridges where viral samples from animals and humans are stored and preserved.

Anthropology of Virus Hunters

The term "microbe hunters" was coined by historian Paul de Kruif in 1926 to describe the progress of microbiological knowledge against the prejudices of superstition. When Leeuwenhoek put flora swabbed from his own mouth under the microscope, argued de Kruif, he invented a rational way to make visible the invisible beings that act on human bodies. The term "virus hunters" was later used during the HIV era, when biologists looked for the origins of the virus in the "wild" forests of Africa (Gallo 1991; McCormick and Fischer-Hoch 1997). The metaphor of hunting was used by "civilized scientists" as a colonial trope to describe the discovery and conquest of a "wild nature." Historians have described how these scientists adopted the perspective of hunters when they collected specimens from members of hunting societies (Anderson 2008; Griffiths 1996). The metaphor of hunting blurs the opposition between those who study and those who are studied, forcing attention to modes of collaboration in the production and accumulation of knowledge. Knowledge production can be described as a form of witchcraft or magic when it negotiates with the invisible. What, then, does the anthropology of hunting societies reveal about the cryopolitics of virology?

Many discussions of biopolitics have turned on its relation to thanatopolitics—making live and letting die (Esposito 2007)—as Foucault opened this question through his analysis of race as the "condition of acceptability of death in a regime that promotes life" (Foucault 1997, 228). But the opposition between "make live" and "let die" (and sometimes make die) remains in the circle of pastoral power, with its mix of sacrifice and surveillance. Recent works have drawn attention to the absence in Foucault's thought of a "cynegetic power" (Chamayou 2010) that takes into account practices of predation that involve the exchange and collection of signs which can lead to the reversal of asymmetries between hunters and hunted. If the production and accumulation of samples of viruses are part of a cryopolitics, it should be understood not only as a problem of "make live and let die" but also as a problem of collecting and conserving forms that circulate between humans and nonhumans, of not letting viruses die so that populations of humans, and in some cases, animals, can be made to live. Microbiologists and virologists, then, behave as hunters not only when they catch viruses in the wild, but also when they store them to observe the continuous mutations underlying the discontinuities between species.

Storage, as a peculiar form of accumulation that does not rely on property, has generated discussion in the anthropology of hunting societies. Alain Testart (1982b) has argued that storage was the first step toward sedentarism in hunter-gatherer societies, as it provides the primary form of wealth. Testart makes a distinction between two kinds of economy: hunter-gatherers who have immediate consumption, such as Australian Aboriginals, and hunter-gatherers who store the results of their collections, such as North-West Coast Amerindians. Storage, he argues, is an accumulation of overproduction to face coming disasters. The first kinds of storage concerned plants and animals and were, in some cases, later extended to raw materials, tools, or pottery. Food storage, argued Testart, brought about a "total change of mentality": the past became more privileged than the present to anticipate the future, and nature came to be seen not as an ever-providing source of sustenance but as a capricious source of disasters.

Storage also brought the first notion of surplus, and without the organization of exploitation in social relationships, the first social inequalities. "In order to be prepared for any eventuality, there will be a tendency to store a little more than the quantity usually needed" (Testart 1982b, 527).

If there is storage in hunter-gatherer societies, it means that the opposition between wild and domestic is not relevant in the description of food practices. Testart contested the opposition made by Gordon Childe between hunter-gatherers and agriculturalists and introduces a third category, "storing hunter-gatherers," which articulated a specific form of surplus on the threshold of domestication. Freezing, Tesart noted, has been one of the modes of storage, besides desiccation, salting, fermentation, or construction of silos, particularly in Arctic hunter-gatherer societies (1982a, 150).

Tim Ingold (1983) has remarked, in response to Testart, that there is no incompatibility between storage and nomadism. Hunter-gatherers can move around a defined territory and yet return to the same points of subsistence. Neither is there incompatibility between storage and sharing in hunting societies. What is stored can be shared in the community, without any recognition of property rights on living things or any form of "appropriation of nature," which for Ingold marks the threshold of domestication. Ingold makes a distinction between practical storage, which is a temporal delaying of the consumption of food resources, and social storage, which involves rights of property and obligations of reciprocity. Social storage subordinates different forms of kinship between elements in stock to the appropriation of nature by society, while practical storage leaves such differences to the ambiguities and paradoxes of practical reason.

This discussion sheds light on the cryopolitics of virus hunters. The problem is indeed to understand how microbiologists can be mobile enough to follow the lines of mutations of strains of viruses and yet store them in localized places without relying on juridical notions of property. While Testart holds an evolutionary view, saying that hunters-gatherers who practice storage introduce inequality before the introduction of agriculture and property, Ingold draws a sharp distinction between hunter-gatherers—characterized by movement and reciprocity—and pastoralists, fixed by sedentariness and property. This controversy, I suggest, comes from the fact that both Testart and Ingold lack a concept of form or information. The most recent innovations in microbiology come from the possibility of storing samples as genetic sequences available through GenBank (an open access resource of all known sequences operated by the US National Institutes of Health). How does the ability to store genetic information change the cryopolitics of viral samples?

The controversy between Testart and Ingold opposes a materialist view of nature as a set of resources and a vitalist view of life as movement in an environment. But the practices of virus hunters today rely on a view of nature as a set of forms (expressed through mutations of genetic sequences) and life as the emergence of new forms (expressed through an infectious event) (Helmreich 2015). The practices of virus hunters are closer to what Paul Rabinow (1996) has called biosociality: among all the forms that can be conceived, those actually giving rise to a collective movement are rare. Claude Lévi-Strauss has described this formal mode of production through the concept of *bricolage*, to which Rabinow refers in his analysis of biosociality. Bricolage is the use of a set of tools or materials that have been assembled randomly, without reference to any project: it is "the contingent result of all the occasions there have been to renew or enrich the stock" (Lévi-Strauss 1966, 17). Even if they are mobile and do not have property rights, hunter-gatherers have a stock of knowledge, as well as resources, that allows them to interpret any event in a pregiven order of things. Lévi-Strauss dedicates long passages of his book, *The Savage Mind*, to stocks of old names that are used for the integration of newborns or new diseases into the collective. Some societies put common names in a reserve where they are prohibited and become proper names ready-for-use after a delay. In the same way, contemporary "virus hunters" show that any new virus emerging among humans, causing unpredictable epidemics, has already been circulating among animals, and that its place can be carved out—or made meaningful—by reference to a preexisting stock of genetic sequences conserved in a fridge.

In the following analysis, I will make a distinction between storage and stockpiling as two modes of conservation of past viruses used to anticipate future epidemics. *Stockpiling* refers to the preservation of vaccines, while *storage* refers to the preservation of samples. When both practices are seen, more fundamentally, as a preservation of forms underlying the distinctions between species, this cryopolitical view sheds new light on biopolitical controversies.

Many of the high-profile discussions about future pandemics have centered on the ethical stakes of stockpiling vaccines (how to distribute them fairly). Vaccines are viruses that are modified, selected, attenuated, adjuvanted, and exchanged. The techniques used to preserve vaccines extend relations of exchange over time. The constitution of vaccine banks relies on the same material techniques as the constitution of virus banks, but the conditions of their preservation differ in meaningful ways. If we start from techniques of preservation rather than from ethical debates, what is the difference between storing samples and stockpiling vaccines regarding modes of anticipation of the future? In the following ethnography, which draws on observations in Hong Kong and Taiwan made between 2007 and 2013, I will point to a slight difference in techniques of freezing that may signal a big difference in cryopolitics. While the storage of samples requires –80°C, stockpiling vaccines is practiced at the relatively higher temperature of 4°C, because vaccines contain adjuvants that boost the immune response. A cryopolitical analysis will elucidate the significance of this difference in degree.

Storing Viral Samples

On April 23, 2013, I visited the Animal Health Research Institute of Taiwan. It is situated on the banks of the river Tamshui north of Taipei, close to a former Dutch military post overseeing the straits between Taiwan and China. The first human case of avian influenza in Taiwan had just been declared. A businessman returning from Shanghai was identified as carrying the H7N9 strain that had infected more than a hundred people in China in the preceding months, killing one out of five. Since the 1997 emergence of the H5N1 virus in Hong Kong, there had not been a human case of avian flu in Taiwan. However, in 2013, the H7N9 virus, although less virulent, seemed to be spreading more rapidly. Despite intense pressures to contain the virus, the head of the Animal Health Research Institute, Hsiang-Jung Tsai, agreed to meet with me and explain practices of surveillance of flu viruses in animals. This was notable especially considering the communication constraints that had been placed on public health officials.

H7N9 is not new, he told me. It had previously been found twice in wild birds in Taiwan, although each time with a different genetic sequence.[5] For the past ten years, the Chinese Wild Bird Society of Taiwan had been collaborating with the Animal Health Research Institute to define areas of study from which they collected bird feces.[6] At the time of my visit in 2013,

50,000 samples had already been collected, out of which 3,000 were found to contain various influenza strains. A monitoring program was also conducted on domestic poultry in farms and markets. In May 2012, around 4,000 cases of H5N2 in poultry farms (not transmissible to humans) were declared to the World Organization for Animal Health (OIE 2012). But there were rumors that the same strain had caused many more cases between December 2012 and February 2013, which had not been declared because of the campaign for the presidential election between December 2012 and February 2013. In July 2012, when smuggled wild birds with H5N1 were found at Taoyuan Airport on a plane coming from Macau, they were immediately destroyed. Even if avian flu had never killed any person from Taiwan, the surveillance of its mutations in animals was a sensitive political question. It revealed tensions in relations between Taiwan and mainland China, since south China was considered since the 1980s to be the epicenter of pandemic flu viruses because of the intense proximity between birds, pigs, and humans (Shortridge and Stuart-Harris 1982).[7]

In this chapter, however, I am not concerned with the identity politics exposed by pathogens when they cross borders and become framed as a security issue (Rollet 2014); nor will I look at the relations between public health and animal health administrations as they try to collaborate despite often diverging economic interests and organizational habits (Keck 2008). Here, I am more interested in the techniques that allow public health experts to formulate a prediction about the virulence of pandemic flu viruses based on the detection of their mutations in birds. The ability of public health officials to claim that a new virus emerging in humans with catastrophic effects was already circulating in birds depended on their ability to study large numbers of accumulated flu viruses from both species. The interface between animals and humans as a site of viral emergence can be ascribed to a relation between the past and the future through techniques of information organization. Future epidemic outbreaks can be traced back to pathogens that circulate silently among animals, in an active form of "latent life" (Radin 2013).

In November 13, 2013, it was reported that a twenty-year-old woman had been infected the previous May by the H6N1 bird flu virus. This was the first recorded instance of that strain having jumped from birds to humans (Kaplan 2013; Wei et al. 2013). The news was announced only in November because researchers from the Taiwan Centers for Disease Control had to first sequence the flu virus of the woman (who had recovered in the meantime) and search analogous flu viruses in samples collected from chickens. The Taiwan Centers for Disease Control ultimately showed that seven of the genes found in the woman were closely related to a flu strain isolated from Taiwanese chickens that year. The eighth gene was most closely related to another strain first found in Taiwanese chickens in 2002. It remains uncertain how this woman came into contact with live chickens, since she was a clerk at a deli, but a "molecular clock" analysis, which tracks the evolution of mutations in a genome, makes it almost certain that the flu she had contracted had previously circulated among birds. In 2013, this knowledge aroused fears that it could jump again to humans, with more catastrophic effects than in 2002.

Let's return to the Animal Health Research Institute. How could you know, I asked Hsiang-Jung Tsai, that it was the same H7N9 that circulated in humans and birds? As an answer, he showed me a fridge, with the temperature −80°C glowing on a red digital thermometer.[8] He told me that 200 wild bird flu viruses were stored in that fridge, two of which were H7N9: for the first, collected in 2009, they sequenced the HA genetic segment, which indicates the virulence, and for the second, collected in 2011, they sequenced eight other segments. Through the efforts of sequencing and comparing mutations in each strain, they could thus show that the 2013 human strain derived from these wild bird strains. The preservation of viral samples in a fridge made possible their synchronization through a phylogenetic tree. As Hannah Landecker writes, freezing techniques, when they were introduced to cultivate cells in the 1950s, synchronized a world of living objects characterized by mutations: "The freezer acted as a central mechanism both within individual laboratories and companies and within the biological research community more generally to standardize and stabilize living research objects that were by their nature in constant flux" (Landecker 2007, 227). Landecker also notes that such practices of low-temperature preservation were used for human cell lines in laboratory work as well as for sperm cells in cattle breeding (157). Synchronization of viral mutations through the use of molecular clocks to construct phylogenetic trees is another instance of the extension of freezing techniques from the lab to the farm and beyond.

I had never been confronted before with the materiality of the techniques by which viruses were stored. I knew that the major virology labs in the influenza world, such as Robert Webster's at St. Jude's Hospital in Memphis, or Peter Palese's lab at Mount Sinai Hospital in New York, or Albert Osterhaus's lab in Erasmus Medical Centre in Rotterdam, had huge collections of flu viruses from humans and animals in their fridges. But the microbiologists I had worked with in Hong Kong didn't show me their fridges, as they were working on computers where they downloaded sequences from GenBank to build phylogenetic trees. One of them, Dhanasekaran Vijaykrishna, described it through the opposition of clean and dirty: "In general, I don't do the lab work, everything that makes your hands dirty. I work mainly on computers. But if I didn't have the staff to do the sequencing, I could do it because that's what I've done for years" (interview with the author in Hong Kong, July 23, 2009). The fridge is used to make dirty feces available as clean sequences through the identification of viruses. It is a technology that can be used to enable the purification of forms of life from the impure materiality of bird dejections.

Maintaining these specimens in the freezer then makes it possible to separate the sequences from their material conditions, allowing them to be used for more abstract evolutionary speculations. Hong Kong microbiologists expressed frustration that most of their samples came from Shantou University, where Hong Kong tycoon Li Ka-Shing had founded a "State Key Laboratory for Infectious Diseases" on the Chinese territory. When Chinese authorities felt the studies of Hong Kong microbiologists threatened their economic interests, they simply forbade them to cross the border, and cut their access to the most recent samples. However, Vijaykrishna claimed: "We're sitting on a bunch of information. ... Viruses are there, still unknown. Even if we don't do surveillance, we have enough information to work for five years." In other words, when cut off from access to older specimens, Hong Kong microbiologists were able to compensate for the scarcity of virus samples by tapping into an abundance of easily accessible digital information. They had turned the event of pandemic emergence into a consistent form, maintaining the laboratory as a "sentinel post" where viral mutations could be analyzed and detected (Keck 2014).

Using this older sequence data, Vijaykrishna and his colleagues were notably able to show that the H1N1 "Spanish flu" virus was already circulating in pigs as early as 1911 (Smith et al. 2009). They didn't need to resurrect the actual virus from some frozen archive: the phylogenic method allowed them to trace contemporary viruses back to "the most recent common ancestor" by probabilistic reasoning.[9] The difficulty of accessing the materiality of viral knowledge, with its "dirty politics" of sovereignty and property, was compensated for by the open access to viral information in the supposedly "clean" relations of trust between scientists on the web.

Stockpiling Vaccines

In another part of the Animal Health Research Institute, fridges contained vaccines for poultry. These were conserved at a temperature of 4°C, which is higher than virus samples, because, I was told, it was sufficiently cold to preserve adjuvants. Adjuvants are proteins that boost the reaction of immune cells. As mentioned above, flu vaccines are live flu viruses cultivated in chicken embryos and attenuated so that they present their antigens to the immune system, which, in turn, produces antibodies. The production of flu vaccines is an industrial challenge, as the flu strain circulating in a population mutates constantly. Every year, influenza experts gather at symposia called by world health administrations (WHO in Geneva for humans, OIE in Paris for animals) to collectively agree on which strains should be the target of vaccination efforts. But for humans, the ordinary production of vaccines against seasonal flu sometimes interferes with the extraordinary production of vaccines against pandemic flu coming from animals.

According to the recommendations of the OIE, vaccination of domestic poultry is a reasonable complement to the slaughtering of infected chickens, but it should also be accompanied by the surveillance of the mutations of flu viruses among wild and domestic birds. While mainland China and Vietnam have been criticized for massively vaccinating their domestic poultry (they were producing these vaccines locally), which could have led to a selection of resistant strains (Smith et al. 2006), Taiwan chose to store vaccines and only use them in case of an outbreak. "We are not supposed to use vaccine, because it interferes with the monitoring by marking the poultry positive," said Hsiang-Jung Tsai. "But if there is an outbreak and the stamping out policy is not enough to stop contagion, we need to use vaccine."[10]

The problem becomes: what do we do with the excess vaccines produced for strains that are no longer circulating? In 2013, the Animal Health Research Institute of Taiwan maintained 10 million doses for H5 and 5 million for H7, the two most currently circulating strains in poultry. They were bought from French (Meyrieux), Italian (Fluvac), and Mexican (Avimex) companies. After 18 months, any unused vaccines are generally incinerated, and updated vaccines are bought. Members of the Taiwanese Parliament complained about the quantity of vaccines destroyed, which led the Taiwanese government to decrease the number of vaccines stockpiled. The government also passed a contract with private companies who are capable of producing 3 million doses of vaccine within a week in case of an outbreak—thus reducing the cost of stockpiling and destroying vaccines. A Taiwanese private pharmaceutical company, Adimmune, announced that it could produce between 5 and 10 million doses of vaccines for H7N9 in six to eight weeks, and was consequently awarded the right to produce the vaccine. It was the first time a Taiwanese company produced human vaccines from scratch instead of repackaging other companies' products (Silver 2013; Chen 2015).

The production of flu vaccines is thus the object of negotiations between public and private interests. While the Animal Health Research Institute could produce and stockpile its own vaccines, because of concerns about waste, it was compelled to rely on the initiative of private companies to invent new technologies of massive and rapid production. Though some saw this as a product of the whims of nature—whether or not an epidemic emerged—others interpreted it as the result of pressure from industry. "We are not supposed to produce too many vaccines—otherwise private companies will not be happy," said Hsiang-Jung Tsai.

Restrictions on public research on vaccines in Taiwan have also been explained as the result of geopolitical constraints: this research aimed to protect the Taiwanese population, and the vaccines could therefore not be exported to any other country. "If we could sell vaccines to mainland China," said Hsiang-Jung Tsai, "public production of vaccines would be profitable, but for now that's not possible. And we are not allowed to buy vaccines from mainland China." It must be recalled that Taiwan had difficulties joining the WHO and OIE because of mainland China's opposition, and has been admitted only in 2009 under the name "Chinese Taipei." The circulation of animal vaccines, or lack thereof, is a material expression of complex geopolitical tensions.

Similar issues are raised by the production of flu vaccines for humans. In 2007, the Indonesian government refused to share samples of flu viruses that circulated in its population—then determined to be the most exposed to H5N1—if it was not granted preferential access to vaccines produced out of them. Minister of Health Siti Fadilah Supari declared in the journal *Nature*:

Samples shared become the property of the WHO collaborating centers in rich countries, where they are used to generate research papers, patents and to commercialize vaccines. But the developing countries that supply the samples do not share in these benefits. In the event of a pandemic, we also risk having no access to vaccines, or having to buy them at prices we cannot afford, despite the fact that the vaccines were developed using our samples. (Butler 2007)

In this case, despite the highly mutable status of flu viruses, the technological possibility of preserving them as samples or even as vaccines raised issues of sovereignty: who was the appropriate party to manage and own the rich biodiversity of flu viruses, recovered from living Indonesian citizens, in order to respond to emerging outbreaks (Lowe 2010)? Since vaccination concerns the health of the whole human population, the possibility of an international biobank of flu strains, where samples would be shared without property rights, was debated.[11]

In June 2009, after the H1N1 virus emerging in Mexico was declared pandemic by the WHO, the French State responded to the potential threat by ordering 95 million doses of vaccines from pharmaceutical companies, most of them European (Sanofi Pasteur and GlaxoSmithKline). This use of the means of production of private companies followed a recommendation of the National Ethics Committee in February, claiming that equal access to vaccines should be guaranteed to the French population in case of a pandemic. As the H1N1 virus turned out to be less lethal than expected, the number of French citizens who received the French vaccination was ultimately low (around 5 million). The French government consequently had to cancel half of its commission of flu vaccines, and created a public organization that would manage the destruction of those vaccines that had been produced but not consumed. It was proposed that former French colonies such as Cambodia or Algeria would receive free vaccines as a kind of drug dumping (Door and Lagarde 2010). The failure of the vaccination campaign can be explained by the fact that the free vaccine was administered in public buildings by requisitioned practitioners, not by private physicians. In France, in contrast with most countries, patients buy their vaccines at the drugstore and keep them in their fridge before bringing it to their physician. The extraordinary stockpiling of pandemic vaccines, with its unusual partnerships between public and private participants, was difficult to understand for ordinary practitioners of vaccine storage.

A last form of stockpiling for flu pandemics concerns antivirals. Used after infection, they inhibit the neuraminidase that controls the entrance of the virus to cells. These antivirals include Oseltamivir (Tamiflu, orally prescribed) and Zanamivir (Relenza, inhaled). In Taiwan, the Department of Health claims it has stocked 20 million doses of Tamiflu. In discussions with Roche, the Swiss company that launched the antiviral ten years ago and became the global provider after 2005, Taiwanese public health officials questioned the company's capacity to produce enough antivirals for the Taiwan population in case of a pandemic. Despite not doing much to promote Tamiflu when it was first launched, Roche gave guarantees that it could deliver them quickly enough.

For vaccines as well as antivirals, the anticipation of a future pandemic gave rise to massive stockpiling and intense negotiations between governments and pharmaceutical companies. While vaccines need to be constantly updated and recycled, antivirals can be kept for five to seven years at a low temperature (2 to 8°C). The capacity to stockpile adequate quantities of vaccines and antivirals is often used as a means of measuring the degree of preparedness of a community for potential coming pandemics. When Hsiang-Jung Tsai was looking at the 200 wild bird strains in the fridge of the Animal Health Research Institute, he said: "Imagine these wild bird viruses could mix with duck farms. There could be direct infection from ducks to humans. We need to be prepared." This enactment—through practices of stockpiling—of an imagined scenario is what constitutes a technique of preparedness.

A Cold War Genealogy: Storing Oil, Stockpiling Weapons

In a series of essays, Andrew Lakoff (2007, 2008) has reflected on preparedness as a contemporary rationality for the anticipation of the future. While insurance captures the future through the calculation of probabilities and risks, he argues, preparedness relies on imagining an event with unknown probability and catastrophic effects. In the ignorance of what this catastrophic event will actually be, techniques have been designed to mitigate all possibilities, from bioterrorist attacks to natural disasters as well as emerging infectious diseases. These techniques combine early warning systems, simulations based on worst-case scenarios, and stockpiling of equipment. Strategies of civil defense during the Cold War—readying the population for a coming catastrophe through simulation exercises—were thus transferred from the threat of nuclear winter to all kinds of hazards. Imagining interactions between wildlife, livestock, and human populations as a viral bomb ready to explode is one of the fictitious horizons, constructed by many scientific novels and reports, that constitutes contemporary preparedness. If "nature is the greatest bioterrorist threat," as microbiologists keep arguing, collecting information about virus mutations is a way to prepare for coming pandemics as an unpredictable event.

Lakoff recalls that the failure of the mass vaccination for the 1976 flu virus, during which 10 percent of the United States population was vaccinated for a flu virus that was less lethal than dreaded, was a shift in public health strategy, paving the way for new techniques of preparedness. If mass vaccination relied on a strategy that covered the whole population for a targeted virus, the side effects of vaccination led to an interruption in the vaccination program. Edwin Kilbourne, a world-famous microbiologist based in New York, who had failed to prevent a flu epidemic in 1947, was one of the most prominent voices to recommend massive vaccination in 1976. He later worked on the genetic engineering of vaccines to mitigate the unpredictability of flu virus mutations, and advocated the production of a vaccine reassorting the thirteen known influenza A virus subtypes, to act as a "barricade vaccine" before a "rampart vaccine" could be made for the actually emerging pandemic virus (Kilbourne 2004). He lived long enough to witness the massive vaccination campaign of 2009, during which his principles were applied.

In the United States, the Strategic National Stockpile is a repository of antibiotics, vaccines, antivirals, and other critical medical equipment. In case of a bioterrorist event or a pandemic outbreak, it will be able to supply local health authorities that may be overwhelmed by the crisis. The basic principle of this kind of stockpiling is that a crisis causes a bottleneck in the distribution of medical equipment that creates a new type of scarcity. Discussions on stockpiling involve models to simulate where bottlenecks will appear in the distribution of equipment. Mapping the distribution system and its potential shortages means reorganizing the population around the vulnerabilities of "vital infrastructures" (Collier and Lakoff 2008). Exercises in clinics lead medical staff to imagine a potential shortage of vaccines. "Who are the most critical actors in case of a pandemic?" is the question generally asked by these scenarios, to justify a seemingly unfair distribution of vaccines and the triage of patients with influenza-like symptoms (Christian et al. 2006). While the catastrophic event has not yet occurred, the logic of preparedness already organizes the classification of objects and persons that will mitigate its effects. Stockpiling has the double meaning of accumulation and classification of objects and persons in relation to a future catastrophe. It is not a totalizing view of the beings that compose a society: it draws a list of essential inventory items identified, procured, and stocked before an imagined event. This fictitious or anticipatory dimension of stockpiling may lead to a comparison with an irrational behavior like hoarding, when wealth is accumulated in such a way that it produces scarcity in the distribution of goods. But stockpiling may also be understood as a rational technique that reorganizes society in the imaginary enactment of a catastrophe.

This analysis sheds light on the distinction between storage and stockpiling as two ways to represent scarcity in the future. Both techniques accumulate goods to anticipate a coming catastrophe, but while storage envisions scarcity at the level of nature, stockpiling simulates it at the level of a society. This point can be endorsed by another Cold War genealogy: that of weapons for a military conflict. In *Carbon Democracy*, Tim Mitchell (2011) shows that the power of Western countries over Middle Eastern states doesn't rely on the abundant sources of oil but, rather, on their capacity to limit them and organize their scarcity through the imagination of shortage. While states relying on carbon as a primary source of energy were always exposed to disruption of their networks by strikes, the regulation of oil states needed the invention of another form of constraint: what Mitchell calls "the economy," an autonomous domain ruled by the laws of exchanges. Western states sold weapons to Middle Eastern countries in return for the massive quantities of oil. "Weapons … could be purchased to be stored up rather than used, and came with their own forms of justification. Under the appropriate doctrines of security, ever-larger acquisitions could be rationalized on the ground that they would make the use of them less likely" (Mitchell 2011, 156). In the context of the Cold War, the stockpiling of weapons was justified by the fact that it forced states not to use them. But in return, it constructed oil as a scarce resource whose economy must be controlled.

An analogy can be drawn here between the storage of oil and the stock-piling of weapons during the Cold War, and, in contemporary biopolitics of infectious diseases, between the storage of viruses and the stockpiling of vaccines. When Indonesia negotiates sending flu strains to control the dis-tribution of vaccines by Western states, it may not be a very different posi-tion from that of oil-producing countries. And the imagination of a future pandemic may act to construct scarcity and increase the value of the viral strains that are exchanged. A natural resource becomes a social resource through the imagination of its shortage. The massive archive of stored viral samples is transformed into a repository of potential weapons through the activation of the immune response, and a potential biobank for the con-flicting property claims on viral information.

Mitchell suggests that the economy of stockpiling weapons follows a logic of "conspicuous waste," whereby what has been accumulated can be destroyed rapidly to show the wealth of a group. This is a reference to Thorstein Veblen's *Theory of the Leisure Class* (1899), where conspicuous consumption and waste is defined as the attribute of an elite that does not follow the rules of work and production. Veblen himself refers to the anal-ysis of *potlatch* among North-American Indians, observed by Franz Boas and later theorized by Georges Bataille and Claude Lévi-Strauss. For Veblen, hunting involved an economy of prestige that was contradictory with the economy of production. Could the massive accumulation and destruc-tion of goods whereby hunting societies show their wealth shed light on regimes of property and valuation in modern societies that accumulate viral strains and vaccines? In that instance, preparedness would not be so new. If it appeared after the Cold War as new, compared to nineteenth-century nation-states' techniques of prevention, it could borrow traits from those hunting societies that had long been studied by social anthropology.

Claude Lévi-Strauss was probably the most influential thinker to con-nect the anthropology of hunting societies with the world of the Cold War in which he lived and thought. His famous distinction between cold and hot societies (Lévi-Strauss 1966, 234) actually points to different modes of storage. Hot societies conceive the future as the unfolding of a mechanism that is already running in the present, on the model of the steam engine or the oil motor: the future expands the past capacities of the machine. Cold societies conceive the future as a confirmation of an order in the past, on the model of the mechanical clock: they "freeze" time, inasmuch as any event already has its place in a past set of relations. As Eduardo Kohn recently noted, "freezing" time in this sense doesn't mean resisting change but rather interpreting human history through a set of forms that connects humans and nonhumans: "What is 'cold' here is not exactly a bounded society. For the forms that confer on Amazonian society this 'cold' characteristic crosses the many boundaries that exist both internal to and beyond human realms" (Kohn 2013, 182). Freezing time, then, is a way to unscale human temporality and rescale it in accordance with the sequences of forms that can be found in the nonhuman realm. Freezing doesn't reduce historical change but enlarges the stock of relations in which it becomes meaningful. Cold societies and hot societies, in Lévi-Strauss's framework, are then both replies to the law of thermodynamics (disorder increases at the expense of order); but while hot societies weave new threads into the social fabric, cold societies use old threads to create a new fabric.

It should be noticed that the famous distinction between hot and cold societies is part of a discussion between Jean-Paul Sartre and Lévi-Strauss on the subject of scarcity. For Sartre, human history—what he calls "dialectical reason"—comes from the consciousness of a scarcity in the resources of existence. He takes as paramount the example of the French Revolution starting with a revolt for bread. For Lévi-Strauss, scarcity is not the product of individual consciousness but a result of collective classifications. In his analysis of the Tiwi system of classification, which he describes as "an inordinate consumption of proper names" (1966, 209), Lévi-Strauss shows that common names are prohibited to become proper names: they are stored and frozen to reintroduce energy in the system, to increase the "semantic charge" (210). This analysis recalls the model of the potlatch that commands the demonstration of Lévi-Strauss's *Elementary Structures of Kinship*: the long series of exchanges of signs, goods, and partners is oriented toward massive destruction, as a potential scarcity built from within the system.

This analysis by Lévi-Strauss sheds light on the distinction between storage and stockpiling for viruses in a contemporary logic of emerging disease preparedness. Storage is the preservation of all kinds of tools and materials, inasmuch as they can be used for different purposes. Its aesthetic value comes from its ordering through the language of forms. Stockpiling is the preservation of samples and vaccines for a pandemic, which orients the production of the pharmaceutical industry. Stockpiling introduces an element of heat in the cold domain of cryopolitics, playing on emotions of panic and inequity, while storage hearkens back to the long duration of relations between humans and animals. This resonates with the fact that stockpiling vaccines is practiced at 4°C, while storage of samples requires –80°C. The presence of adjuvants in vaccines introduces a live element that triggers the reaction of the immune system, while samples can be preserved for their antigenic information. It should be added that in case of a pandemic, the refrigerating system adds another layer of vulnerability to the infrastructure: it is better not to rely on freezing if the infrastructure collapses in the course of a pandemic. While storage creates forms out of any type of scarcity, stockpiling simulates a scarcity to orient its production of forms.

I have considered storage and stockpiling as two modes of production of biovalue (Waldby 2002; Rajan 2006). Virus strains become a source of profit when there is speculation on their value for the future. Among all the viruses that circulate in animals, we make a bet on the strain that will become pandemic if it jumps to humans. The genetic sequence of that strain can be submitted for a patent if a profitable vaccine can be made out of it. However, I have removed the analysis of this accumulation of strains from the framework of pharmaceutical patenting, in order to look at a form of accumulation that relies not on property but on an "ostentatious" logic, which is the practice of collectors. This accumulation is speculative not in the sense of a financial bet on future profits but in the sense of an intense surveillance of present relationships (the first meaning of the Latin *speculare* is "watching over"). The collection and preservation techniques of microbiologists may be closer to those of highly skilled hunters, who follow the lines of animal movements and trace their relations of kinship, than to those of pharmaceutical companies.

The anthropology of hunting societies thus illuminates the contemporary production of biovalue in a cryopolitical regime. If any technique of anticipation of the future produces value through the construction of scarcity, stockpiling meets the demand in a standardized way while storage follows the logic of a collection that needs to be completed. Every sample has its own value, but there are gaps in surveillance that can be filled through molecular sequencing. If stockpiling is close to hoarding, as a form of accumulation that produces destruction by the imagination of scarcity, storage is closer to collecting, as a practice of accumulation oriented toward the aesthetic pleasure of sharing. These distinctions between different techniques of accumulation and anticipation can be refined further by making distinctions between different sites of conservation: stores, stocks, repositories, banks, archives, or registries. The cryopolitics of viral samples circulating between humans and animals allows us to draw distinctions in the world of biobanks and biovalue. These distinctions, as they are produced by techniques of freezing intended to make live and not let die, are a matter of degree in the variation of production temperature, and not epistemological ruptures between ontologies of nature.

Notes

1. Geographer and anthropologist Kirsty Duncan set up a similar expedition in Longyearbyen, Norway, but after months of obtaining authorization to excavate and thaw the bodies of flu victims, it turned out that the samples were not viable as the bodies were not in the permafrost (Duncan 2003).

2. The names of flu viruses come from the hemagglutinin (H) and neuraminidase (N) proteins that control the entrance and release of viruses in cells. Numbers depend on their chronological order of identification.

3. Interestingly, this is a commonly held explanation for the severity of the 1918 flu, which also disproportionately killed the young and healthy. See Kolata 2005.

4. The number of live chickens consumed in Hong Kong in 2002 was 30 million and the number of chilled chickens almost zero. In 2008, the number of chilled chickens consumed was 35 million, and the number of live chickens was 5 million (source: Hong Kong Food and Environment Hygiene Department).

5. A viral strain is classified by the antigenic information it produces. However, the genetic sequence may reveal differences in the same viral strain, as some of the genetic information is not relevant to the life cycle of the virus. But these differences reveal phylogenic mutations of the same viral strain.

6. On the history of bird societies in Hong Kong and Taiwan, and their relations with environmental health authorities, see Keck 2015.

7. There is an obvious colonial history of considering China as the epicenter of emerging infectious diseases and its margins (Japan, Taiwan, Hong Kong, or Singapore) as sentinel territories for the spread of these viruses to the rest of

the world. However, this is also a Cold War genealogy, as China and Taiwan have been potential enemies, and rumors about bioterrorism abound between the two countries.

8. Flu virus samples from birds are stored at 4°C for two weeks after collection, which is the time for virus identification and molecular sequencing, then at –20°C for rapid transport, and then at –80°C (Munster et al. 2009).

9. The article of Smith et al. (2006) begins by contesting Taubenberger's 2005 hypothesis that the Spanish flu may have jumped directly from birds to humans, which would explain its lethality. This interpretation, they argue, is controversial because of variant gene phylogenies that either conflict with this theory or remain ambiguous because of a lack of contemporaneous viruses. The phylogenic method allows them to fill this gap.

10. If a vaccinated chicken tests positive, it is impossible to know whether the antigens come from exposure to the flu virus or from the injection of the vaccine.

11. The WHO had recommended an international surveillance of flu viruses since the 1950s, with a collective sharing of the various strains. However, what was new with the "emerging diseases worldview" (King 2002) was the idea that strains emerging in the South should be exchanged for vaccines produced in the North.

References

Anderson, Warwick. 2008. *The Collectors of Lost Souls: Turning Kuru Scientists into Whitemen*. Baltimore: The Johns Hopkins University Press.

Butler, Declan. 2007. Q&A: Siti Fadilah Supari. *Nature* 450 (7173): 1137.

Caduff, Carlo. 2012. The semiotics of security: Infectious disease research and the biopolitics of informational bodies in the United States. *Cultural Anthropology* 27 (2): 333–357.

Caduff, Carlo. 2014. Pandemic prophecy, or How to have faith in reason. *Current Anthropology* 55 (3): 296–315.

Chamayou, Grégoire. 2010. *Les Chasses à l'homme*. Paris: La Fabrique.

Christian, Michael D., Laura Hawryluck, Randy S. Wax, Tim Cook, Neil M. Lazar, Margaret S. Herridge, Matthew P. Muller, Douglas R. Gowans, Wendy Fortier, and Frederick M. Burkle. 2006. Development of a triage protocol for critical care during an influenza pandemic. *Canadian Medical Association Journal* 175 (11): 1377–1381.

Collier, Stephen, and Andrew Lakoff. 2008. The vulnerability of vital systems: How "critical infrastructure" became a security problem. In *The Politics of Securing the Homeland: Critical Infrastructure, Risk, and Securitisation*, ed. Myriam Dunn and Kristian Kristensen. London: Routledge.

Door, Jean-Pierre, and Jean-Christophe Lagarde. 2010. *Rapport sur la manière dont a été programmée, expliquée et gérée la campagne de vaccination contre la grippe A(H1N1)*. Paris: Assemblée Nationale.

Duncan, Kirsty. 2003. *Hunting the 1918 Flu: One Scientist's Search for a Killer Virus*. Toronto: University of Toronto Press.

Esposito, Roberto. 2007. *Bios: Biopolitics and Philosophy*. Minneapolis: University of Minnesota Press.

Foucault, Michel. 1997. *Il faut défendre la société*. Paris: Gallimard-Seuil.

Gallo, Robert. 1991. *Virus Hunting: AIDS, Cancer and the Human Retrovirus: A Story of Scientific Discovery*. New York: Basic Books.

Greger, Michael. 2006. *Bird Flu: A Virus of Our Own Hatching*. New York: Lantern Books.

Griffiths, Tom. 1996. *Hunters and Collectors: The Antiquarian Imagination in Australia*. Cambridge: Cambridge University Press.

Ingold, Tim. 1983. The significance of storage in hunting societies. *Man* 18 (3): 553–571.

Kaplan, Karen. 2013. Taiwanese woman is the first human to be sickened by H6N1 bird flu. *Los Angeles Times*, November 13.

Keck, Frédéric. 2008. From mad cow disease to bird flu: Transformations of food safety in France. In *Biosecurity Interventions: Global Health and Security in Question*, ed. Andrew Lakoff and Stephen Collier, 195–225. New York: SSRC-University of Columbia Press.

Keck, Frédéric. 2014. From purgatory to sentinel: "Forms/events" in the field of zoonoses. *Cambridge Anthropology* 32 (1): 47–61.

Keck, Frédéric. 2015. Sentinels of the environment: Birdwatchers in Taiwan and Hong Kong. *China Perspectives* 2: 41–50.

Kilbourne, Edwin. 2004. Influenza pandemics: Can we prepare for the unpredictable? *Viral Immunology* 17 (3): 350–357.

King, Nicholas. 2002. Security, disease, commerce: Ideologies of postcolonial global health. *Social Studies of Science* 32 (5–6): 763–789.

Kohn, Eduardo. 2013. *How Forests Think: Toward an Anthropology beyond the Human*. Berkeley: University of California Press.

Kolata, Gina. 2005. *Flu: The Story of the Great Influenza Pandemic of 1918 and the Search for the Virus That Caused It*. New York: Simon & Schuster.

Lakoff, Andrew. 2007. Preparing for the next emergency. *Public Culture* 19 (2): 247–271.

Lakoff, Andrew. 2008. The generic biothreat, or How we became unprepared. *Cultural Anthropology* 23 (3): 399–428.

Landecker, H. 2007. *Culturing Life: How Cells Became Technologies*. Cambridge, MA: Harvard University Press.

Lévi-Strauss, Claude. 1966. *The Savage Mind*. London: Weidenfeldt & Nicolson.

Lowe, Celia. 2010. Viral clouds: Becoming H5N1 in Indonesia. *Cultural Anthropology* 25 (4): 625–649.

McCormick, Joseph B., and Susan Fischer-Hoch. 1997. *The Virus Hunters: Dispatches from the Frontline*. London: Bloomsbury.

Mitchell, Timothy. 2011. *Carbon Democracy: Political Power in the Age of Oil*. New York: Verso.

Munster, Vincent J., Chantal Bass, Pascal Lexmond, Theo M. Bestebroer, Judith Guldemeester, Walter E. P. Beyer, Emmie de Wit, et al. 2009. Practical considerations for high-throughput influenza a virus surveillance studies of wild birds by use of molecular diagnostic tests. *Journal of Clinical Microbiology* 47 (3): 666–673.

OIE. 2012. *Six-Monthly Report on the Notification of the Presence of OIE-Listed Diseases*. http://www.oie.int/wahis_2/public/wahid.php/Reviewreport/semestrial/review?year=2012&semester=1&wild=0&country=TWN&this_country_code=TWN&detailed=1 (accessed September 30, 2014).

Peiris, Joseph, Menno de Jong, and Yi Guan. 2007. Avian influenza virus (H5N1): A threat to human health. *Clinical Microbiology Reviews* 20 (2): 243–267.

Rabinow, Paul. 1996. Artificiality and enlightenment: From Sociobiology to biosociality. In *Essays on the Anthropology of Reason*, 91–111. Princeton, NJ: Princeton University Press.

Radin, Joanna. 2013. Latent life: Concepts and practices of human tissue preservation in the International Biological Program. *Social Studies of Science* 43 (4): 483–508.

Radin, Joanna. 2014. Unfolding epidemiological stories: How the WHO made frozen blood into a flexible resource for the future. *Studies in History and Philosophy of Biological and Biomedical Sciences, Section C* 47: 62–73.

Rajan, Kaushik Sunder. 2006. *Biocapital: The Constitution of Postgenomic Life*. Durham, NC: Duke University Press.

Rollet, Vincent. 2014. Framing SARS and H5N1 as an issue of national security in Taiwan: Process, motivations, and consequences. *Extrême-Orient Extrême-Occident* 2:141–170.

Shortridge, Kennedy, and Charles Stuart-Harris. 1982. An influenza epicentre? *Lancet* 320 (8302): 812–813.

Silver, Dave. 2013. Tiny Taiwan preps for worst; H7N9 vaccine plan in place. *BioWorld Today* 24 (71): 1.

Smith, G. J. D., X. H. Fan, J. Wang, K. S. Li, K. Qin, J. X. Zhang, D. Vijaykrishna, et al. 2006. Emergence and predominance of an H5N1 influenza variant in China. *Proceedings of the National Academy of Sciences of the United States of America* 103 (45): 16936–16941.

Smith, Gavin, Justin Bahl, Dhanasekaran Vijaykrishna, Jinxia Zhang, Leo L. M. Poon, Honglin Chen, Robert G. Webster, J. S. Malik Peiris, and Yi Guan. 2009. Dating the emergence of pandemic influenza viruses. *Proceedings of the National Academy of Sciences of the United States of America* 106 (28): 11712.

Sung-Hsi, Wei, Ji-Rong Yang, Ho-Sheng Wu, Ming-Chuan Chang, Jen-Shiou Lin, Chi-Yung Lin, Yu-Lun Liu, et al. 2013. Human infection with avian influenza A H6N1 virus: An epidemiological analysis. *Lancet* 1 (10): 771–778.

Taubenberger, Jeffery K., Ann H. Reid, Amy E. Krafft, Karen E. Bijwaard, and Thomas G. Fanning. 1997. Initial genetic characterization of the 1918 "Spanish" influenza virus. *Science* 275 (5307): 1793–1796.

Testart, Alain. 1982a. *Les chasseurs-cueilleurs ou l'origine des inégalités*. Paris: Société d'Ethnographie.

Testart, Alain. 1982b. The significance of food storage among hunter-gatherers: Residence patterns, population densities, and social inequalities. *Current Anthropology* 23 (5): 523–537.

Veblen, Thorstein. 1899. *Theory of the Leisure Class*. New York: Macmillan.

Waldby, Catherine. 2002. Stem cells, tissue cultures, and the production of biovalue. *Health* 6 (3): 305–323.

Freezing Ontologies

7 Reflections on the Zone of the Incomplete

Deborah Bird Rose

Anthropocene

The age of humankind: this is the era in which human action has become a planetary force. We know the climate change issue well because it has the greatest profile, but it is only one element of a much wider set of entwined processes that include the great mass extinction event now in process, the acidification of the oceans, the accumulation of plastic waste, the loss of soils and fertility, the loss of rainforests, and the rampant consumption that fuels the work of tearing up and wrecking the earth.[1]

In the words of the biologist E. O. Wilson, our species is a planetary killer (Wilson 2002, 79). Many scientists now hold the view that the earth has passed a "tipping point" from which there is no return to anything like earth as humans have experienced it (McKibben 2014; see also Lynas 2007). It is possible that our species too will go extinct. If so, it will not happen because we have reached the end of our life span. All species, with a few exceptions, have predictable life spans; the "problem" with the current mass extinction event is that so many species are being cut short so suddenly. The end for us humans, like that of so many other species, is likely to take place because directly and indirectly we have made life on earth impossible.

This is a vast uncontrolled experiment, as Baudrillard explained in his book *The Illusion of the End*, and it is not confined to humans:

Man is without prejudice: he is using himself as a guinea-pig, just as he is using the rest of the world, animate and inanimate. He is cheerfully gambling with the destiny of his own species as he is with that of all the others. ... He cannot be accused of being a superior egoism. He is sacrificing himself, as a species, to an unknown experimental fate. (Baudrillard 1994, 83)

In the shadow of present and future catastrophe, cryotechnologies raise enticing possibilities of continuity across disaster and on into some future, more welcoming world. The techno-optimism of these efforts is enmeshed in complex Western end-time thinking, so my aim is to explore constructs concerning the end time as both a temporal moment bounded by a "before" and "after" and as an extinction event. Extinction is loss; it is an end with no return. And yet, technologies for freezing biotic fragments, freezing whole bodies, and, even more radically, engineering the "resurrection" or re-creation of dead species seek to overcome extinction—to treat it as a reversible or salvageable condition. Necessarily they work with an imagined end time and beyond. They draw some of their cultural power from Western dreams of stopping, kick-starting, or leaping across time, and from modernity's commitment to achieving eternal perfection here in the world of actual beginnings and endings. These techno-utopian dreams work in a zone of the incomplete—a culturally imagined time-space between the edge of the end that is coming and the edge beyond that end (McKibben 2003, 157). This zone is relevant to individual lives, to the lives of species, and to histories, cultures, and eons. A whiff of redemption pervades all thought concerning this strange zone.

Disruptions

For decades now, environmental philosophers, activists, and many others have developed a strong critical analysis of a Western epistemology that is deeply implicated in setting the Anthropocene crisis in motion, particularly the narrative that offered visions of progress and held forth the hope that technology would engineer the world and deliver a happy ending. The Anthropocene, in the wide sense that includes both the monumental global changes now in process and the recognition of humanity's role in contributing to these changes, is now doing what decades of insightful critique never quite succeeded in doing: it is forcing the truth upon us. The Anthropocene is something of a mirror, and its reflection of human agency is grotesque—an agency that outstrips its capacity to manage itself; that wrecks, pillages, loots, and destroys; that has very little idea what it is doing; and that carries with it, in contradiction to all reason, an expectation of immunity. It always needs to be said, of course, that both causal responsibility for the looming catastrophe and vulnerability to it are unequally distributed among members of the human species as well as across other species.

Almost all of my research these days involves extinctions, and I am acutely aware of living in a time of imminent and extremely unhappy endings. To think of endings is to invoke time, and in this context, too, the Anthropocene brings Western thought into radical disarray. The Anthropocene is now unmaking Western hubris, which asserted that whatever catastrophes might be encountered from time to time, none had irreversible consequences (Neithammer and Templer 1989, 32).

This topic is taken up in an exemplary way in Nick Mansfield's (2008) essay, "'There Is a Spectre Haunting …': Ghosts, Their Bodies, Some Philosophers, a Novel, and the Cultural Politics of Climate Change." Mansfield's diagnosis and its prognosis are both dire. Working with Derrida's (1994) ideas about hauntology—in which ideas about the immediacy of the present are supplanted by "the figure of the ghost as that which is neither present, nor absent, neither dead nor alive"—Mansfield presses us toward the understanding that we are facing a catastrophic unmaking with irreversible consequences. It arises out of our own past, and comes at us from our future. Neither familiar nor totally unfamiliar, it is destroying our sense of history and freedom, as Chakrabarty (2008) and others have also discussed. Mansfield (2008) shows that the Anthropocene wrecks the logic of both modern and postmodern sensibilities; it is not that their logic and meaning are coming undone, but more that they "are about to hit a wall, and that wall … cannot be made friendly, [and it] is climate change." It is an "Absolute Other who abolishes and does not make new." In Mansfield's analysis, "the material violence of the past emerges, reincarnate, re-fleshed, in our future, and in a politics for which our last centuries of politics cannot prepare or even forewarn us." As the Anthropocene destabilizes and appears to overturn Western concepts of history, progress, and a future waiting to be filled according to human choice, it reveals how nonlinear, how truly convoluted, these constructs actually are. Nowhere is this more legible than in the fact that technologies of cryopreservation rely implicitly on end-time thinking involving what I call "the zone of the incomplete."

Time Ideology and Linearity

As is well known, the dominant modern Western concept of time starts with a "homogenous general magnitude" (Apffel-Marglin 2011, 152–154). There is a roll-out of time in which the portion that is past is being filled up and the portion that is future is empty time waiting to be filled. Michelle Bastian analyzes clock time and the out-of-sync modes of living that it masks. She is particularly interested in coordinated universal time, explaining that clock culture emphasizes "continuity and similarity across all moments, and projects an empty and unending future" (Bastian 2012, 33). Within this construct, time is unilinear, monological, and teleological. The unilinear aspect is a one-directional, future-facing direction of history. It is monological in the sense of claiming to be true for all people on earth, sooner or later. It will scoop them all up in a vision that ingeniously combines engineering and freedom, and will deliver *its* vision of the good life to all (eventually).[2]

The story underlying the secular version of this concept of time is progress: from simple to complex, and from uniformity to diversity, along with movement toward other desiderata that provide an account that puts Western humans at the top of the human ladder, and puts humans at the top of a ladder of evolutionary life. The triumph of Western humans in the evolutionary stakes is attributed to "intelligence" or some equally self-congratulatory term.

Although the coordinated universal time program works with the idea of an unending future, Western time constructs hold fast to an eschatological view of human progress, asserting that time has a direction leading toward a known outcome: always more progress, more of the good life, and (sometime/somewhere) utopia. All kinds of violence and misery can be, and are, justified as necessary steps toward the promised future.

History, in this time ideology, is an exercise of power deployed in the present, defining the past, and aiming toward the future. Those who get in the way will be jettisoned. Thus, along with its claims to universality and the good life, this ideology also has a far more violent side, perhaps best acknowledged in Hegel's term "slaughterhouse of history." The relentless determinacy that powers "history" is a ruthless exercise in exonerating those who deal in death by claiming that they are fulfilling a historical mission.[3]

In this register, then, history is a secularized version of messianic time. The substitution of religious thought with historical thought based on reason that was accomplished in the nineteenth century replaces one myth with another. In Neithammer and Templer's words, both myths worked with "the same mold": "history was conceptualised as moving toward a final endpoint whose quality was oriented in terms of the earlier hopes that had been pinned on a world beyond the world" (Neithammer and Templer 1989, 34).

Like "progress," messianic time, too, is unilinear, monological, and teleological—heading toward the end of time that will be heralded by the return of the Messiah. The violence of history is mirrored in the apocalyptic violence of the messianic return. Particularly in its Christian iteration, the end of the history arrives with terrible violence and entails the end of many lives, not just in the moment, but in eternity. The future, in this apocalyptic vision, belongs to some but not all. Universality is part of the story in this way: there is no escape from the violence of this future that is coming toward us. There is only at best the prospect of survival and making it through to utopia.[4]

The sense of coming catastrophe signaled by the term "Anthropocene" is calling forth decisions that affect huge numbers of species and ecosystems, including humans. For example, widening the zone of the incomplete to include all that cryotechnology and other technologically optimistic interventions can come up with makes it necessary to choose whose body parts are included in the Frozen Ark. As I have indicated, the whiff of redemption is rarely far from this zone.

The Elasticity of Time and the Long Transitive Moment

In both secular and messianic forms, this Western time construct is curiously elastic and filled with irregular and repetitive stops and starts. The end point is always just at the edge of the horizon.

Recent studies such as Bastian's defamiliarize this mainstream story by showing the convolutions and peculiarities that the big story of time's one-way arrow slides across. It depends on a stable imagery—years, days, atomic clocks—and yet, all that is being measured is only one kind of time. The imagined certainty of knowing *the* time obscures all the ways in which we are out of sync with other times (Bastian 2012, 35). For the moment, though, I want to stay with irregularities and elasticity. Not only can the Western time story be stretched to cover everything, it can be punctuated to form a dizzying array of new beginnings, endings, periods, post-periods, and eras (think "Anthropocene"). In its future orientation, present problems will be solved, and the horizon of resolution always glimmers at the edge of "now."

I have written elsewhere about the year zero and the long transitive moment in the context of the frontier (Rose 2004). The year zero is the moment of encounter leading to conquest—a time punctuation that asserts that an old era has ended and a new one begun. The long transitive moment is the stretched time between the year zero and the actual arrival of the desiderata (civilization, for example) that was proclaimed as the rationale and destiny of the new era. And yet, with wonderful elasticity, the end point is still (always) arriving. It is imagined to pronounce the final shape of meaning to the whole. But as the story posits an end that hasn't yet arrived, life, meaning, and even death may be suspended in a zone of the incomplete.[5]

The story as it can be discerned in areas deemed to be frontiers is exemplary of a pattern that is one of the key characteristics of modernity. We are close to the horizon, but not (or never?) quite there yet. The nation is always being built, technologies are always being improved; there is perpetual progress, but never arrival at perfection. On the one hand, the idea of motion through time toward the end corresponds neatly with the image of linearity. But on the other, the teleology of the story masks this curious effect of elasticity and indeterminacy. In the politics of time and narrative, while we are waiting for the end, everything seems to be up for grabs.

The end of the story—which has not yet been reached—is the moment of ultimate meaning. The linearity of time that is oriented toward the end thus becomes complicated by the idea that when the end arrives, it will wash back over the time that preceded it, and remake this time. It is helpful to think of this reversal of history's meanings in terms of the Messiah. He will arrive, and he will sort out the histories. He will separate the winners from the losers, reclaim the lives and stories of the winners, and make it all meaningful. In the messianic vision, death, along with all suffering and injustice, will be overcome for some but not all. It is the ultimate and final moment of decision making about who lives and who dies.

Restoration and Apocalypse

The teleology of history thus works with a toxic combination of hope directed toward the distant future and willful disregard of suffering, even willful implementation of death-work in the present and near future.[6] Susan Handelman addresses Jewish time constructs as well some Christian visions. Drawing particularly on the work of Gershom Scholem and Walter Benjamin, she describes two forms of hope directed toward the distant future or "end of history": restorative and apocalyptic (Handelman 1991). A restorative vision seeks to achieve utopia through restoration of a more perfect past. Hope works toward the past, drawing inspiration from the idea that there has been utopia here on earth and can be again. A Christian iteration of the restorative vision is the return to Eden, analyzed admirably by Carolyn Merchant. An apocalyptic vision looks toward the future, knowing that violence and suffering will be required as humanity leaps toward a utopia that has never had an earthly counterpart. It allows for no smooth transitions, and at the same time is profoundly hopeful of redemption (Merchant 2003).

The darkest scenarios of what the effects of the many great, uncontrollable, world-changing processes now in process (the Anthropocene) are likely to mean for life on earth involve an apocalypse that holds no hope of redemption. Indeed, the Anthropocene wrecks both visions: the restorative vision is wrecked because everything will be changed deeply and permanently, and the utopian vision is wrecked because there is no reason at all to expect a perfected future.

Walter Benjamin expressed just such wreckage in his "Angel of History" essay. Here the past has been obliterated and there will be no restoration. Here the obliteration of the past leads directly to the violence of history from which there is no redemption. This beautiful essay is perfect prophecy. In one paragraph Benjamin gives us the unmaking of hope, the power of catastrophe, and the powerlessness of redemptive action. He tells the truth not only of his time and place but of a coming time and place that is ours, too:

A Klee painting named *Angelus Novus* shows an angel looking as though he is about to move away from something he is fixedly contemplating. His eyes are staring, his mouth is open, his wings are spread. This is how one pictures the angel of history. His face is turned toward the past. Where we perceive a chain of events, he sees one single catastrophe which keeps piling wreckage upon wreckage and hurls it in front of his feet. The angel would like to stay, awaken the dead, and make whole what has been smashed. But a storm is blowing from Paradise; it has got caught in his wings with such violence that the angel can no longer close them. This storm irresistibly propels him into the future, to which his back is turned, while the pile of debris before him grows skyward. This storm is what we call progress. (Benjamin 1969, 257–258)

Messiah Envy

The Western history of conquest and transformation can be read as a series of mini-Messiah moments—year zero end-times when many are cast into eternal death (extinctions), and a few are allowed to thrive and prosper.

In our day, we are seeing yet another mini-Messiah moment in relation to plants and animals at the edge of extinction. There is a technology-driven view that much is still salvageable. Species, it is postulated, are something like broken toys—bio-engineering will fix or reassemble them. The earth's climate system is like a huge machine that is malfunctioning—geo-engineering will get it working properly again. Stewart Brand's book of warning asserts: "We are as gods and HAVE to get good at it" (Brand 2009).

Brand is wrong, of course. We are nothing like gods. He does, however, perfectly replicate the desire to be a Messiah. This is the desire to arrive in the midst of catastrophe and decide who will be saved and who will be cast out. Technologies of cryonics and de-extinction work with this mode of messianic favor. De-extinction goes the furthest by playing with ideas of turning time around and running it backward, resurrecting the dead and scooping up individuals near extinction to ensure that species extinction never quite arrives. All the actions that cause extinction continue in the world. These technologies are end-time oriented; they move some matter into a zone of suspended life—enlarging the zone of the incomplete—in order to be able to kick-start time and life again when the moment arrives.

This project, then, claims already to be remaking the past by undoing death. It continues the war of winners and losers by extending the transitive moment of the incomplete by pulling dead species back into life. Or, to recall other projects—to freeze tissue samples for a future day when all will be well, the promised future will have arrived, and species can live again. The result is, in effect, to freeze time by holding lives, body parts, and samples within the zone of the incomplete. In contrast to the slogan "Kill them all and let God sort them out," cryotechnologies pose a different one: "Save them all. Let future Scientists sort them out."[7]

An astonishing hope lurks in these technologies. They depend on creating a weirdly optimistic time-space zone that stops time and holds open the possibility that either restoration or utopia may still be possible. Even more radically, they hold the paradoxical hope that both may be possible.

This techno-apocalyptic vision involves a recombinant mingling of restoration and utopia. That which has lived and died will live again, and a mingling of species and ecosystems that never were will fulfill a vision of an even better world. Scientists will preside Messiah-like over the final distribution of life and eternal death.

After the End

Western time constructs have depended not only on the idea of an end time but on a redemptive post-end—a time and a place from which the winners of history can tell the story, and enjoy the fruits, of their victory. The wreckage of the Anthropocene is not yet fully upon us, but this "irresistible storm" already is washing back over the hopes and visions that fueled the West's sense of destiny (Benjamin 1969, 257). The search for the technological fix is a scramble to rescue the post-end, to claim that redemption is still possible. And yet it is probable that as the Anthropocene continues to unmake those dreams and hopes, it is bringing us into growing recognition of an impossible possibility, an "unimaginable end" (Mansfield 2008).

The problem of imagining an end that actually is an end entails ethics, as James Hatley contends in a complex and fascinating article on what ethics and virtue could mean in our current condition of "wilful decreation" (Hatley 2012, 20). He is contemplating the ethical dimensions of learning to accept and affirm endings, and to turn our life action toward the mutual co-flourishing of life here on earth. In this Levinas-inspired account of virtue, care for others in our present moment and place is the basis for repairing the ruins of this earth.

Notes

1. See Levene 2013 for an excellent review of the evidence on which my summation is based, and for a discussion of the implausibility of recovery.

2. This vision depends on a dream of immunity that discounts all the species, societies, ecosystems, languages, and ways of life that are being utterly extinguished.

3. Fabian (1983) has offered a hugely insightful analysis of time, colonization, and the role of anthropology.

4. My concern is with the structure and moral valence of this vision, so I am not exploring its many permutations, one of which is the fascinating idea of the Rapture, by which certain people (and probably only humans) will be allowed to escape the suffering of the end times.

5. Some Christian religions hold that the human dead are suspended, awaiting resurrection.

6. The ongoing colonization of the world offers ample evidence of this assertion.

7. I associate this slogan with the Vietnam War, but it may have a much older origin in the crusades against the Cathars (see: http://www.thisdayinquotes.com/2011/07/kill-them-all-and-let-god-sort-them-out.html).

References

Apffel-Marglin, Frédérique. 2011. *Subversive Spiritualities: How Rituals Enact the World.* Oxford: Oxford University Press.

Bastian, Michelle. 2012. Fatally confused: Telling the time in the midst of ecological crises. *Environmental Philosophy* IX (1): 23–48.

Baudrillard, Jean. 1994. *The Illusion of the End.* Cambridge: Polity Press.

Benjamin, Walter. 1969. *Illuminations.* Trans. Harry Zohn. Ed. and with an introduction by Hannah Arendt. New York: Schocken Books.

Brand, Stewart. 2009. *Whole Earth Discipline: An Ecopragmatist Manifesto.* New York: Viking.

Chakrabarty, Dipesh. 2008. The climate of history: Four theses. *Critical Inquiry* 35 (winter): 197–222.

Derrida, Jacques. 1994. *Specters of Marx: The State of the Debt, the Work of Mourning, and the New International.* New York: Routledge.

Fabian, Johannes. 1983. *Time and the Other: How Anthropology Makes Its Object.* New York: Columbia University Press.

Handelman, Susan. 1991. *Fragments of Redemption.* Bloomington: Indiana University Press.

Hatley, James. 2012. The virtue of temporal discernment: Rethinking the extent and coherence of the good in a time of mass species extinction. *Environmental Philosophy* 9 (1): 1–22.

Levene, Mark. 2013. Climate blues: Or How awareness of the human end might re-instil ethical purpose to the writing of history. *Environmental Humanities* 2:147–167.

Lynas, Mark. 2007. *Six Degrees: Our Future on a Hotter Planet*. London: Harper Perennial.

Mansfield, Nick. 2008. "There is a spectre haunting ...": Ghosts, their bodies, some philosophers, a novel, and the cultural politics of climate change. *borderlands e-journal* 7 (1).

McKibben, Bill. 2003. *Enough: Staying Human in an Engineered Age*. New York: Times Books.

McKibben, Bill. 2014. Climate: Will we lose the endgame? *New York Review of Books* 61 (12): 46–48.

Merchant, Carolyn. 2003. *Reinventing Eden: The Fate of Nature in Western Culture*. New York: Routledge.

Neithammer, Lutz, and Bill Templer. 1989. Afterthoughts on posthistoire. *History & Memory* 1 (1): 27–53.

Rose, Deborah. 2004. *Reports from a Wild Country: Ethics for Decolonisation*. Sydney: UNSW Press.

Wilson, Edward O. 2002. *The Future of Life*. New York: Alfred Knopf.

8   Out of the Glacier into the Freezer: Ötzi the Iceman's Disruptive Timings, Spacings, and Mobilities

David Turnbull

On September 19, 1991, a Swiss couple walking in the Ötzal Alps near the Austro-Italian border found a man's body emerging from the melting ice. He became famous as "Ötzi the Iceman," the oldest human to be found fully preserved complete with all his belongings. The ongoing extensive forensic examination of this remarkable example of cryopreservation provides an opportunity to focus on the disruptions and irruptions that freezing technologies appear to produce in biological and social spatiotemporal processes, revealing a complex interweaving of time, space, mobilities, technologies, and politics.

Much of what has been taken for granted about time—the questions of its irreversibility, its linearity, and its unity—has been based in assumptions about temporality that underpin contemporary physics. Likewise, space and spatiality have largely been abstracted and neutralized and rendered nonproblematic in biological processes. In recent years questions of temporality, spatiality, and mobility have come to be rethought in a radical intersection between biology, physics, geography, linguistics, anthropology, and science studies.

For instance, the geographer Doreen Massey (2005) insists that politics is not possible without seeing both space and time as multiple and open-ended. Bruno Latour (1997) argues for dispensing with the terms "time" and "space" as naming stable entities in favor of thinking of events, actions, and processes as products of timing and spacing. Lee Smolin (2013) argues for the reintroduction of time to physics. Susan Oyama and much recent work on complex adaptive systems and epigenetics suggest that varieties of temporality and spatiality are central to understanding developmental processes (Oyama, Griffiths, and Gray 2001). These spatiotemporal threads have become rewoven in the recognition that spatiality and temporality are constructed in social processes and embodied movement (Ingold 2000, 2011; Turnbull 2007).

The constructedness, interrelatedness, and plurality of space and time are becoming recognized in the social sciences. However, we still lack a fully articulated framing that can account for the continuous transformations of space and time within changing sociotechnical formations like globalization and communications technologies. In the biological sciences there is an emerging physicochemical narrative explaining timing and spacing in terms of neurotransmitters and chemical signaling (Kholodenko et al. 2010). But, there is also an emerging nonphysicalist biological narrative of complexity, mentioned earlier, that makes the lens of cryopreservation especially valuable in examining reconfigurations of both time and space, and the reconstitution of domains that have been sliced into "physics" and "biology" or even "natural" and "cultural."

To bring spatiotemporalities and cryopolitics together, I want to start by looking at examples of emergence—emergence, that is, of fully fleshed ancient bodies. Since the seventeenth century, hundreds of bodies have been found in bogs mainly in northeast Europe—Cashel Man being the most recent and also the oldest dated at around 2000 BC (McGrath 2013). Ancient ice bodies are much rarer. Kwäday Dän Ts'ìnchi (meaning "Long Ago Person Found"), a 500-year-old Tlingit man, emerged in 1999 from a British Columbia glacier (Dickson and Mudie 2008; CBC News 2008; Plicht et al. 2004). Kwäday Dän Ts'ìnchi's body was examined along with only a few possessions that remained intact. He carried a knife with trade metal blade, and his gut contents suggest he was moving inland, across glaciers, from the coast where he probably lived, and encountering white traders. In her brilliant book *Do Glaciers Listen?* Julie Cruikshank (2005) describes just such encounters between the knowledge traditions of the Tlinglit and Western traders and glaciologists. She points out that for Western scientists the behavior of local glaciers was mysterious and erratic, as they are subject to sudden and catastrophic movement. For the Tlinglit, glaciers have agency and must be treated with respect. It is their intimate knowledge of the ice and the glaciers that enable the Tlingit to orient themselves and move through an otherwise dangerous and unpredictable landscape.

However, the most exceptional example of cryopreservation—an ancient human body almost totally preserved right down to his DNA and gut microbes—is Ötzi the Iceman. He was frozen in ice for 5,300 years before he emerged in 1991 at the top of the Italian Alps. After he was found he was transferred to a freezer in the South Tyrolean Museum of Archaeology in Bolzano, where he will remain in sociotechnical suspension as long as the power supply is sustained and processes of degeneration can be held at bay.[1] Ötzi, like Kwäday Dän Ts'ìnchi, was frozen very shortly after death, but two important differences separate them. Unlike Kwäday Dän Ts'ìnchi's, all Ötzi's belongings were preserved, and a much more sustained analysis of his body is permissible owing to his great antiquity and the lack of any continuity in the ethnographic record. Because there is a continuous living tradition, Kwäday Dän Ts'ìnchi's body was not subject to extensive scientific examination, his remains were treated according to Tlingit tradition, and his body may not be displayed. Surprisingly, both Kwäday Dän Ts'ìnchi and Ötzi have identifiable living relatives revealed by DNA analysis of mitochondrial DNA haplotypes, but in Kwäday Dän Ts'ìnchi's case contemporary Tlingit can identify him as a member of the Wolf/Eagle moeity or clan who lived near salt water and had shared in the Tlingit ice knowledge tradition that along with his adaptive clothing gave them the mobility to cross the glaciers inland (Champagne and Aishihik First Nations 2009; Dickson and Mudie 2008). Unfortunately for science, Ötzi's ice knowledge and how he moved and oriented himself remain frozen, though his material technologies reveal much about his ability to extend himself in space and time.

In the case of Ötzi, what differences do the circumstances of his cryopreservation make in his move from the glacier to the freezer? How does he differ from, say, a frozen mammoth or even an ice core? These are questions that open up the ways in which Ötzi is deeply imbricated in multiple temporalities and spatialities, both before his death and after his cryopreservation. The forms of disruption, dislocation, discontinuity, and disorientation that accompany changing technologies of assemblage, orientation, and movement create differing narratives of space and time. Unavoidably, to talk about Ötzi is to create a narrative ordering events in space and time. The recursive difficulty is compounded by the reality that from the first moment of the discovery, for the frozen corpse to take on the identity of "Ötzi the Iceman" meant being enmeshed in complex and competing narratives of spatiality and temporality.

Ötzi's body was found partly emerged from the ice on the high alpine Tisen Pass very close to the Italian-Austrian Border. Two questions quickly became the focus of scientific attention: How old was he and where exactly was he located? Answering these would determine who was responsible for him and the relevant expertise necessary to make sense of his remains. The copper axe beside him suggested this was not a modern body, but one that may be 4,000 years old, and hence of profound archeological concern. The question of where he was turned out to be much more problematic, heavily inflected by the conditions of the ice in which he was preserved, the melting of which led to his emergence. Though he was acknowledged to have been close to the border between present-day Austria and Italy, he was first assumed to be Austrian, hence his Austrian name, and the body was sent to the University of Innsbruck where he spent six years under careful examination. But owing to the thick ice cover, the border had only been loosely marked by distantly separated cairns. The question of Ötzi's retrospective nationalization was so intense that it led to an official surveying and marking of the border. It was determined that Ötzi was spatially located 92 meters on the Italian side of the border. After intense negotiations he was rechristened with an Italian identity and transferred to the Museum of the South Tyrol in Bolzano. However, the story did not end there. The Treaty of St. Germain in 1919 that redistributed the territories of the Austro-Hungarian Empire following the end of World War I ceded the South Tyrol to Italy and defined the border formally and naturalistically as the main watershed of the Alps separating the new reduced Austria to the north from the augmented Italy to the south. The topographic "reality" of that watershed and Ötzi's location was revealed when the meltwater from his excavation was found to drain away to the north into Austria (Kutschera and Müller 2003). This was a topographic reality incompatible with political reality in which Ötzi resides in an Italian museum. As a cryopolitical subject, Ötzi's existence undermines ideas of linear historical narratives and discrete national borders. His emergence reveals new spatiotemporal dimensions of geopolitics within which narratives of origins and belonging are constructed and negotiated.

Ötzi is now acknowledged as the oldest preserved human body, with the oldest extant blood cells (Pappas 2012). What makes Ötzi's body the subject of such intense scientific scrutiny is his remarkable state of preservation. Not only is his body almost completely preserved right down to his DNA, his blood cells, and gut contents, so also are the clothes and tools found surrounding his body. It is the revelation of such intimate personal details of an individual who lived in a distant and otherwise barely discernible past that is so temporally disorienting, and so demanding of narrative incorporation and attribution of identity, agency, and mobility (Lynch 2008).

What has been gleaned from Ötzi's body? It is widely agreed that he died 5,300 years ago. He was five feet and five inches or 165 centimeters tall; he had a few fleas, intestinal whipworm, a little arthritis and diarrhea, and he may have been infertile (Hardach 2006). He lacked a twelfth rib and two other ribs were healed having been broken earlier in life. Computer tomography reveals major arterial calcification that, along with gall stones and periodontitis, indicates a surprisingly rich and starchy diet (Seiler et al. 2013). He had considerable wear on the upper left jaw because he probably used his teeth as tools to process wood, leather, bones, and sinew (South Tyrol Museum of Archaeology 2013). Whole genome sequencing in 2010 showed he had type O blood and brown eyes, was lactose intolerant, and had an ancestry in common with inhabitants of Corsica and Sardinia with an unusually high 5.5 percent Neanderthal component (Keller et al. 2012). Ötzi's Y chromosome analysis shows he is in a rare haplogroup G-L91, and nineteen living relatives, members of this haplogroup, have been identified. This genomic data ties them to Ötzi across time and space to a "particular geographical location—the Öztal Alps in this case—and early migratory routes" (Draxler 2013).

His occupation, however, has always been a matter of keen speculation and contestation, and central to disputes over his identity. The Bolzano Museum is keen to locate him as an exemplar of Germanic folkways. Hudson et al. (2011) have suggested the concept of *occupationscape*, that is, "landscapes formed and performed through histories of occupational behavior," as the form of narrative spatialization that has been deployed to ethnically reincorporate Ötzi into the South Tyrol: "After the South Tyrol was ceded to Italy following the First World War, occupations in this region have been used to negotiate ethnicity through an idealized contrast between the rural, agricultural lifestyles of the German-speaking and the more urban, craft- and industry-focused activities of Italian-speaking populations." This fits with his portrayal by some as a shepherd involved in transhumance. However, he had no woolen or woven clothing, no shepherd's equipment, and a recently revised view holds that transhumance did not begin in the region till around 1500 BC (Oeggl, Dickson, and

Bortenschalger 2000; Science Daily 2011). Others suggest he may have been a herder since they have been able to identify some of the leather in his leggings and coat as coming from sheep, and in his moccasins from cattle (Hollemeyer et al. 2008). Some claim he was a copper worker, since he carried a copper-headed axe and his hair had traces of arsenic, copper nickel, and manganese (Gossler et al. 1995), but it's not yet clear if the arsenic was deposited by bacteria after death, or whether it was in his hair while he was living. He may have been a hunter, but there is some possibility that he also had some special status, given that his clothes were so meticulously sewn and his rare and unusual axe was probably very valuable (Dickson, Oeggl, and Handley 2003). Still others claim he was a trader. Though he carried no trade goods, many of his materials came from distant sites. Yet he was not on a trade route when he died, being close to, though well above, the Neolithic trade route passes. Much speculation has been focused on him being a shaman since he carried a medicine bag containing leather thongs on which were threaded two pieces of a common birch fungus *Piptoporus betulinus*, which contains polyporic acid C, an effective antibody, especially against stomach microbacteria. This indicates, if not a specific knowledge of herbs and natural ingredients, at least a general acceptance of their use.

He was, however, most definitely a wayfarer, as indicted by the proportions of his leg bones and pelvis, the wear and tear of his knees (Ruff et al. 2006). And certainly he displayed a highly developed capacity for movement across a complex and difficult terrain, as is shown by his specialized equipment from a variety of different sources. By implication, he must have had the same degree of intimate knowledge of the ice possessed by Kwäday Dän Ts'ìnchi (Skeates 2013). It has been regarded as unusual that he reached the age of about forty-five, a very considerable age for a Neolithic man. However, he died extremely violently in a fight with at least three other people upon whom he also inflicted some damage. He stabbed one with his dagger, two of his arrows were broken, but had been retrieved after he had shot two people with them. His own body shows extensive bruising, defensive cuts to his hands, and an arrowhead under his shoulder that severed an artery and may ultimately have killed him, though he also had cranial bruising indicating a blow to the head (Lorenzi 2007). Blood on the shoulder of his cape suggests he either carried a wounded companion, or was himself carried by someone, who also probably turned him over to withdraw the arrow that may have killed him, which could account for the unnatural posture of his body (MacIntyre 2003).

Debate is ongoing about whether the disposition of his body and his belongings are evidence of burial rather than abandonment. Using spatial point pattern analysis, rather than focusing on the corpse alone, Vanzetti's Italian research team precisely mapped the position of the entire assemblage. Their modeling allowed for the transport of the materials by ice over time and led them to conclude that the spatial positionings were not random but deliberate, as in that of grave goods (Vanzetti et al. 2010). Oeggl and his Austrian team disagree, arguing that, for example, the upright position of the bow is not likely in a burial; and so the debate continues (Zink et al. 2011).

It is now agreed that his death at 3,210 meters near the Tisen Pass occurred in late spring. Furthermore, where he was born, where he lived, and what he did in the last thirty-three hours of his life have now been established by some remarkable track following.

Ötzi's origins and life have been revealed in part by the imprint on his teeth and bones of the "unusual geological and topographical complexity" of where he lived; something that wouldn't have been possible if he had lived on the relatively featureless plains of Iowa, for example (Holden 2003). The isotopes of oxygen in the rain differ on the two sides of the Alps, as a consequence of their differing sources in the Atlantic and Mediterranean. Ötzi's tooth enamel, which was laid down in early childhood, shows the southern pattern and hence that he was born on the Mediterranean side of the Alps. In addition, his bones show the isotope ratios of four different rock types that are only found in that combination in a very few valleys of the South Tyrol. Wolfgang Müller at the Australian National University and his team of geologists used the two types of isotope data to establish his most likely place of birth as the village of Feldthurns where a megalith from the same era has been found (2003). But Ötzi had at some point ingested twelve minute grains of mica, presumably from a grindstone used in preparing his last meal of wheat and barley. Those grains were dated at ages consistent with rocks in a small area in the Etsch Valley. The Müller group concluded that he grew up in the Eisack Valley and spent his adult life further west—a conclusion that fits with another bit of forensic spatial tracking.

The analysis of his gut contents tracks the path of three meals he had in the estimated thirty-three hours before he died. On the day before he died, he ingested, along with his meal, pollen from pine trees growing at around 2,500 meters in a subalpine valley east or west of the Schnals Valley. From there he descended to what was probably his home village, where Juval Castle now is at the junction of the Schnals and Etsch. This is apparent because about ten hours before he died he had another meal that also contained pollen, but this time from the hop-hornbeam typical of the broad-leafed valley floor trees, along with the mica from local grinding stone. Four to seven hours later he climbed back into a subalpine coniferous forest where he had another meal again marked by pine pollen. The combined results of botanists, histologists, and anatomists make it possible to create what could be called a "gastro-intestinal chronotopograph," a gut map charting a path linking the timings and spacings of Ötzi's movements in the Etsch Valley through mapping the varieties of biolocatable materials in his digestive system. When Ötzi was briefly defrosted in Bolzano, gut content samples were taken from a sequence of sites in his gastrointestinal tract. Both a rectal sample, and a sample from the end of the large intestine, the sigmoid colon, had high levels of pine and spruce pollen, indicating a subalpine meal thirty-three hours before death. A second higher site in the transverse colon had high levels of pollen from hop-hornbeam, hazel, and birch, typical of lower valley broad-leaf trees, indicating a meal ten hours before death. The last two sites in the transverse colon, and the ileum terminalis at the end of the small bowel, showed high pine and spruce pollen content marking his penultimate meal at subalpine levels around three hours before death. His stomach went unnoticed and unexamined until recently, owing to its displacement into the normal position of the left lung. Its contents show that 30 to 120 minutes before he died in the Tisen Pass, Ötzi ate a final meal of ibex with an admixture of hair and flies (Oeggl et al. 2007; Gostner et al. 2011).[2]

The fine mapping of the details of Ötzi's life right through from birth down to his last days, hours, and even minutes is revealed through the complex intersections of a formidable variety of technologies and disciplines. From geology to microbiology, from "atomic force" microscopy to cosmology, they all serve to locate Ötzi in space and time. However, a contradiction is concealed in the apparent unity. What is omitted in creating a unified narrative of Ötzi's life is that the ontological assumptions and infrastructures sustaining these technologies and disciplines are not independent of scale. They are not all singing from the same spatiotemporal song sheet. They operate from a wide variety of incommensurable scale-dependent measures of space and time (Gosden and Kirsanow 2006). For example, the isotope decay measures used to determine his diet at given ages operate on a different timescale from those used to determine the age of bones or what he had to eat in his final hours: "Ötzi highlights how complex the temporality surrounding an individual can be. Any person represents a point of intersection of varying temporal scales" (ibid.). The work, the mediation, and negotiations required to achieve a contingent and temporary spatiotemporal ordering before disorder reemerges are not apparent in the narrative; they are invisible (Star 1990). Instead, a seamless space/time is simply presumed to preexist, allowing the appearance of a unified narrative (Jones et al. 2004).

What Ötzi's artifactual assemblage unequivocally shows is his surprising and impressive capacity for mobility. He moved with evident ease and rapidity over considerable heights and distances in extremely difficult and highly variable terrain and conditions. This raises the question of how he was able to achieve such mobility and autonomy. To extend himself in space and time Ötzi had to have the social and technological capacities of connectivity, communication, and orientation. To move, he needed to make connections, material and social. As the archaeologist Robin Skeates (2013) has shown, Ötzi did not act as a lone individual; he was connected across space and time to other "dividuals." But his mobility also required material connection. Ötzi had to have intimate knowledge of the landscape and of the capacities, or what Gibson (1979) called "affordances," of a wide variety of materials and techniques (mostly locally specific), and of ways of making them mobile, adaptable, and repairable, fundamentally using string (Turnbull 2008).

Ötzi was very well dressed and expertly equipped, using materials that had been very carefully selected as best suited to their function, including some nonlocal materials like flint and copper. He had three layers of specially treated skins and grasses enabling him to operate effectively in extreme alpine environments. He had a belt from which to drape his loincloth and suspend his leggings, a jacket, and a bearskin hat—all made of small pieces meticulously sewn using fine stitching and thread, suggesting a specialized craft industry, contrasting strongly with the rough grass stitching of what were probably his own running repairs.

He had shoes ingeniously made of string and calf, deer, and bear skin, and stuffed with grass for warmth. So ingenious in fact they have now been reproduced and are set to go into commercial production in the Czech Republic for climbers who find them ideal for rough terrain (Hall 2005).

His gear included a longbow staff made of yew which he was in the process of shaping and a birch quiver full of viburnum (wayfarer tree) arrows; a yew-handled copper axe; a sheathed dagger with an ash handle; a larch- and hazel-framed backpack (though possibly this was part of a pair of snow shoes); two birch bark containers, one containing charcoal; and a belt pouch housing small useful items including flints for fire making, a retouching tool, and fungus for tinder. He was tattooed with 73 simple designs consisting of 13 groups of parallel lines and 3 crosses possibly with shamanic or medical implications (Sjovold et al. 1995).

But what is most important, though not often focused on, is the large amount of binding material found with Ötzi and his equipment, in the form of cords and cord fragments, and even pieces of plaited material. Typically he used the best available lime bark for a range of tasks; sometimes just bast strips, roughly separated inner bark fibers. For other tasks the fibers were retted and more refined (Pfeifer and Oeggl 2000). He carried a string net, possibly for bird trapping. His cape of vertically arranged grasses was knotted together with horizontal twines. The scabbard for the dagger was a masterpiece of bast plaiting (Spindler 1994, 105–138). The quiver contained a coarse rolled string, and he carried lots of spare string materials with him, though he had no woven clothing. Everything he wore and used required sinew, grass, or bast string; without the technology of connection that string provided he couldn't travel. He was thus as much a "string man" as he was an "iceman."

Ötzi, like the scientific experts examining his corpse, worked with and across a wide and incommensurable variety of spatial and temporal scales using a complex mix of heterogeneous practices. Ötzi was extraordinarily self-sufficient; he was able to create and sustain his own personal mobile microclimate with his own clothes and tools. Yet at the same time his technologies of mobility, such as his axe and finely sewn clothing, were most likely produced by other expert and specialized crafts people. This suggests that he was both independent and embedded in a system of exchange, though it's not clear what his role in it would have been (Hodder 1999, 137–145; 2000, 21–33). The point being that as we struggle to make sense of Ötzi's life through narrative reconstruction, we are stuck not only with the necessary tension of working with macroscale, long-term, grand narratives of human behavior that are not commensurable with microscale narratives of lived moments of individual lives, but also with the multiplicity of spatial and temporal scales that the plethora of analytic disciplines brings to the story. But so too was Ötzi; through his technologies of combination and coordination, he created both topological and cognitive paths in his travels through the landscape. Yet somehow the tensions and contradictions of his own independence and his relationship to a wider community led to his being killed. Ötzi had agency in the active performance of his life just as we do now in the retelling of it. His frozen corpse seems to be an exemplary singular material body—a unique and special, specifically localized and static individual. But his lived body emerges as multiple, performatively sustained in complex evolving webs of semiotic normalizations, technological infrastructures, and agentive recursive coproductions, interwoven with diverse temporalities and spatialities (Mol 2002; Butler 1993; Hayles 1999; Haraway 2008).

How then to locate ourselves and Ötzi temporally, spatially, and narratologically in this teeming multiplicity? One way Ötzi can be located is at the heart of a background debate over how we designate the current era—geologically/naturally as the Holocene, or naturally/culturally as the Anthropocene (Stromberg 2013). Geologists are wedded to the seemingly clearly demarcated Holocene, the period that began at the end of the ice age and which can be given specific temporal boundary at 11,700 years ago. Others argue that humans and their activities are a critical factor not only in climate change, but in all global processes. Accordingly, we should acknowledge that we are in the human-inflected Anthropocene. However, there is no agreement on when to designate its inception—the Neolithic revolution, the industrial revolution, or the atomic age? From my perspective, and if we take Latour's (1993) and Haraway's (2008) insights seriously that we have never been modern nor human, we need to go a lot further back to recognize that humans have always coproduced themselves in interaction with their technologies and with other species. And as a consequence humans have continuously affected what might otherwise be seen as purely natural processes. Once the nature/culture dichotomy is undone,

one has to go back at least to the invention of tools, string, and fire and the adoption of dogs as companions. I would argue for going back to at least two million years ago, which would predate *Homo sapiens*. Arguably the invention and production of stone tools went hand in glove with the development of language and the brain, but in turn the development of the brain was dependent on the invention of fire and cooking enabling the shortening of the digestive tract and the release of energy allowing for cognitive development (Dunbar et al. 2010; Wrangham 2009; Arsuaga 2002). The connective sociotechnologies of string, stories, and social networks served to augment communication through external symbolization in the form of bead necklaces and pendants and made possible the mobility that enabled humans to occupy every micro-niche on the planet (Barber 1994; Bednarik 2008). In my view, what makes Ötzi significant—what makes him cryopolitical—is that he is the oldest fully preserved example we have of this performative coproduction of knowledge and space in action.

Finally, what we need is some spatiotemporal grounding in the apparent tension between a globalized singular space coupled with a unidirectional neutral time against the plethora of spatialities and temporalities—the block universe and the unreality of time versus the vagaries of relativity and contingency. The tension between these dichotomies is more a result of narrative framing than a verifiable or absolute reality. It's often claimed that physicists either ignore time or stick to a simple linear sequence, but in fact, depending on their focus, they deploy whatever version of time best suits their explanation. The standard thermodynamics model naturally assumes the unidirectional linear "arrow of time," since so many physical events are irreversible, entropy increases, and order decreases. In the Newtonian universe, temporal reversibility is the norm; physical laws are time independent and work forward and backward. In the Einsteinian universe, space and time are one; in quantum mechanics, "time is sometimes two-dimensional, sometimes reversible … and at other times discontinuous and fractal-like" (MIT Technology Review 2010). The cosmologist Lee Smolin (2013) has added a controversial twist, arguing that the laws of physics are not timeless but can evolve over time.

Thermodynamics and views like Smolin's suggest that physics and biology have a profound temporal commonality in historicity and contingency. Darwin's evolutionary theory opened biology time up to time. This provided, as Liz Grosz (2005) points out, "a new ontology, a new understanding of life, not as enduring stable essentials, but as a process of emergent becoming, something with a temporal dynamic in continuous unpredictable interaction with its material surrounds. ... Darwin introduced an indeterminacy into life, it's unstable, open-ended, no goal nor predictability." This temporal opening up led to developmental systems theory, ecology, and the recognition that organisms have agency and variation—they interact with each other and their environments transform them in the process. Time and space are not neutral coordinates within which this happens; they too are transformed (Kwa 2010).

The spatiotemporal narratives underpinning "evo-devo" and complex adaptive systems contrasted strongly with those of contemporary genetics where information is the building block of material reality, information that is carried by genes (von Baeyer 2004). In the purely genetic information world, "biological function depends on spatio-temporally organised signaling networks" (Kholodenko et al. 2010). While the informational metaphor currently dominates, some biologists are keen to point out that biological processes are profoundly topological: they have to do with shape, position, transformation, and interaction. For Honig and Rohs (2011), protein–DNA interactions are better seen as "the binding of two large macromolecules that have complex shapes and considerable conformational flexibility." Travis (2011) argues that this is also the case at the next level up from genetic activity, observing that the cell has structure and localization. In other words, where proteins are in the cell and where they are in relation to each other has an effect on activity. Things have to be in the right place at the right time.

However, a substantive issue lies at the heart of biology, genetics, and neuroscience: there is no commonly accepted account of the emergence of life out of inorganic matter, or consciousness out of electrochemical processes in the brain. Arguably the problem is that the informational metaphor has tended to crowd out other perspectives such as spatiotemporal emergence. Paul Griffiths (2001) rightly points out that genes serve to organize biological processes in time and space, but the inherent temporality and spatiality of such processes is obscured by thinking simply in terms of information.

Another area of biological research requiring the development of fresh spatiotemporal narratives is the recently revealed microbial world in which human genetic material is less than 10 percent of the total occupying the body, the rest belonging largely to microbes. According to Nathan Wolfe (2012), 40 percent of the genetic material in the human body is biological "dark matter" consisting of forms of life that are not yet recognized or classified. It could be like *Archaea*, a whole kingdom of new forms we simply have not yet contemplated. Identity and genes, species, evolution, natural kinds, and biological causality are now thoroughly and profoundly disrupted. It is no longer enough to simply think of human beings as uniquely genetically identified individuals; it also needs to be recognized that they can also be conceived in a variety of ways, including as emergent agents assemblages of mobile, interacting, cooperating communities of symbionts.

Biology thus has at least two different spatiotemporal narratives running in parallel. In the neo-Darwinist genetic lineage narrative, genes are units of inheritance—coded instructions for making an individual. The remainder of the genetic material is simply additional complexity which can be accumulated as data to be mined in support of the already articulated genetic reality—a reality that has a spatiality and temporality that can be mapped, explained, and modeled, only slightly simplistically, by the tree of life. The tree of life is obviously a spatiotemporal narrative laying out in space and time speciation events as branches of the tree with its unitary origin point from which all life forms developed, and the objects of evolutionary change are defined in terms of spatial boundedness (i.e., species, organisms, individuals, or genes).

But the recognition of a plethora of alternative processes in the genotype/phenotype relationship such as horizontal or lateral gene transfer challenges the spatiotemporal narrative underpinning the linear sequentiality of the tree of life, and serves to open up a whole new spatiality and temporality previsioned by Deleuze and Guattari as rhizomatics, actor network theory as a network, and Maturana and Varela as a web (Thompson 2007). All of which aim to dispense with the examination of events in terms of physical and chemical interactions of material bits of stuff, and replace it with an interactionist account where things and events are the effects of actions and processes in interaction creating systems of connection that in turn have emergent effects that feed backward into the system reinforcing or undoing it (Grosz 2005; Dupré 2012; Oyama et al. 2001). These performative emergent systems-based accounts sustain very different narratives of spatiality and temporality. Time is not simply unidirectional or effectless. Biological events have historicity; they are contingent and causal, changing the system not only in, across, and against time, but in differing cycles and scales. So timings, rather than time, are important dynamics. Space likewise is not simply a neutral container in which spatially bounded entities meet, mate, and part, but is both a consequence of the movements of the agents in the system, and a dynamic in the interactions in that system, so spacings, like timings, are key (Latour 1997). Movement, orientation, connectivity, and coordination become central to explanation and the making of order and meaning. Spacing and timing are thus contested and negotiated, requiring, as Latour argues, constant work to sustain them (Jones et al. 2004).

Ötzi's ability to move, make connections, orient, and coordinate his life in a multiplicity of spacings and timings was dependent on constant communication, negotiation, and interaction, and likewise his continued existence and our deepening understandings of him and his cultural context are also suspended in emergent and multiple spacings and timings. So what is the difference between his life encapsulated in the glacier and in the freezer? The answer would seem to be that as the plethora of writers have attempted to suspend Ötzi in space and time by telling a unified frozen narrative with an attendant set of preservation practices, he resists fixation and freezing, constantly reemerging in a complex evolving process. So maybe there is no difference between being in the ice and in the freezer, except for the commitment to preservation and analysis that his emergence and transference brings into being.

Notes

1. Ötzi is kept in a special chamber at −6.12°C and at 99.42 percent humidity. His first display chamber had inadequate humidity levels and he lost 5 grams of water every 24 hours. There is a backup chamber in case of breakdown. X-rays show some gray spots are appearing on his knee possibly caused by bacteria produced gas under the skin, but the invasive procedures needed to determine this are no longer allowed. See Deem 2012.

2. Interestingly, a very similar chronotopograph could be drawn tracking the last days of the Alaska ice man. See Dickson et al. 2004.

References

Arsuaga, Juan Luis. 2002. The Neanderthal's necklace. In *Search of the First Thinkers*. New York: Four Walls Eight Windows.

Barber, Elizabeth W. 1994. *Women's Work: The First 20,000 Years*. New York: W. W. Norton.

Bednarik, Robert G. 2008. The mythical moderns. *Journal of World Prehistory* 21:85–102.

Butler, Judith. 1993. *Bodies That Matter: On the Discursive Limits of Sex*. London: Routledge.

CBC News. 2008. Scientists find 17 living relatives of "Iceman" discovered in B.C. glacier. *CBC News*. http://www.cbc.ca/news/canada/british-columbia/scientists-find -17-living-relatives-of-iceman-discovered-in-b-c-glacier-1.761267 (accessed September 30, 2014).

Champagne and Aishihik First Nations. 2009. Kwaday Dan Ts'inchi long ago person found. *Champagne and Aishihik First Nations Special Report*. http://cafn.ca/wp -content/uploads/2015/04/febMar2009newsletter1.pdf (accessed August 25, 2016).

Cruikshank, Julie. 2005. *Do Glaciers Listen? Local Knowledge, Colonial Encounters, and Social Imagination*. Vancouver: UBC Press.

Deem, James M. 2012. His icy chamber. *James M. Deem's Mummy Tombs*. http://www.mummytombs.com/otzi/chamber.html (accessed August 25, 2016).

Dickson, James H. 1995. *Ötzi's Last Journey: Plants and the Iceman*. Glasgow: Glasgow University Press.

Dickson, James H., Klauss Oeggl, and Linda Handley. 2003. The Iceman reconsidered. *Scientific American* 288 (5): 60–69.

Dickson, James H., Michael P. Richards, Richard J. Hebda, Petra J. Mudie, Owen Beattie, Susan Ramsay, Nancy J. Turner, et al. 2004. Kwäday Dän Ts'ìnchì, the first ancient body of a man from a North American glacier: Reconstructing his last days by intestinal and biomolecular analyses. *Holocene* 14 (4): 481–486.

Dickson, James H., and Petra J. Mudie. 2008. The life and death of Kwaday Dan Ts'inchi, an ancient frozen body from British Columbia: Clues from remains of plants and animals. *Northern Review* 28:27–50.

Draxler, Breanna. 2013. Living relatives of Ötzi the Iceman found in Austria. *Discover Magazine Blog*. http://blogs.discovermagazine.com/d-brief/2013/10/16/ living-relatives-of-otzi-the-iceman-mummy-found-in-austria/ (accessed September 30, 2014).

Dunbar, Robin, Clive Gamble, and John Gowlett. 2010. The social brain and its distributed mind. In *Social Brain, Distributed Mind*, ed. Robin Dunbar, Clive Gamble, and John Gowlett, 3–13. Oxford: Oxford University Press.

Dupré, John. 2012. *Processes of Life: Essays in the Philosophy of Biology*. Oxford: Oxford University Press.

Gibson, James J. 1979. *The Ecological Approach to Visual Perception*. Boston: Houghton Mifflin.

Gosden, Chris, and Karola Kirsanow. 2006. Timescales. In *Confronting Scale in Archaeology: Issues of Theory and Practice*, ed. Gary Lock and Brian Molyneaux, 27–38. New York: Springer.

Gossler, Walter, Claudia Schlagenhaufen, Kurt Irgolic, Maria Teschler-Nicola, and Hrald Wilfing. 1995. Priest, hunter, alpine shepherd, or smelter worker? In *The Man in the Ice*. vol. 2. Ed. K. Spindler, E. Rastbichler, H. Wilfing, D. z. Nedden, and H. Nothdurfer, 269–273. Vienna: Springer-Verlag.

Gostner, Paul, Patrizia Pernter, Giampietro Bonatti, Angela Graefen, and Albert R. Zink. 2011. New radiological insights into the life and death of the Tyrolean Iceman. *Journal of Archaeological Science* 38 (12): 3425–3431.

Griffiths, Paul E. 2001. Genetic information: A metaphor in search of a theory. *Philosophy of Science* 68 (3): 394–412.

Grosz, Elizabeth. 2005. *Time Travels: Feminism, Nature, Power*. Durham, NC: Duke University Press.

Hall, Allan. 2005. Czechs cobble new line in prehistoric footwear, as worn by the Iceman. *Age*, July 18.

Haraway, Donna J. 2008. *When Species Meet*. Minneapolis: University of Minnesota Press.

Hardach, Sophie. 2006. Ötzi the Iceman may have been infertile. *ABC Science*. February 6. http://www.abc.net.au/science/articles/2006/02/06/1562968.htm (accessed September 30, 2014).

Hayles, N. Katherine. 1999. *How We Became Posthuman: Virtual Bodies in Cybernetics, Literature and Informatics*. Chicago: University of Chicago Press.

Hodder, Ian. 1999. *The Archaeological Process: An Introduction*. Oxford: Blackwell.

Hodder, Ian. 2000. Agency and individuals in long-term processes. In *Agency in Archaeology*, ed. Marcia-Anne Dobres and John Robb. London: Routledge.

Holden, Constance. 2003. Forensic geochemistry: Isotopic data pinpoint Iceman's origins. *Science* 302 (5646): 759–761.

Hollemeyer, Klaus, Wolfgang Altmeyer, Elmar Heinzle, and Christian Pitra. 2008. Species Identification of Oetzi's clothing with matrix-assisted laser desorption/ ionization time-of-flight mass spectrometry based on peptide pattern similarities of hair digests. *Rapid Communications in Mass Spectrometry* 22 (18): 2751–2767.

Honig, Barry, and Remo Rohs. 2011. Biophysics: Flipping Watson and Crick. *Nature* 470:472–473.

Hudson, Mark J., Mami Aoyama, Mark C. Diab, and Hiroshi Aoyama. 2011. The South Tyrol as occupationscape: Occupation, landscape, and ethnicity in a European border zone. *Journal of Occupational Science* 18 (1): 21–35.

Ingold, Tim. 2000. *The Perception of the Environment: Essays in Livelihood, Dwelling, and Skill.* London: Routledge.

Ingold, Tim. 2011. *Being Alive: Essays on Movement, Knowledge, and Description.* London: Routledge.

Jones, Geoff, Christine McLean, and Paolo Quattrone. 2004. Spacing and timing. *Organization* 11 (6): 723–741.

Keller, Andreas, Angela Graefen, Markus Ball, Mark Matzas, Valesca Boisguerin, Frank Maixner, Petra Leidinger, et al. 2012. New insights into the Tyrolean Iceman's origin and phenotype as inferred by whole-genome sequencing. *Nature Communications* 3 (2): 1–9.

Kholodenko, Boris N., John F. Hancock, and Walter Kolch. 2010. Signalling ballet in space and time. *Nature Reviews: Molecular Cell Biology* 11 (6): 414–426.

Kutschera, Walter, and Wolfgang Müller. 2003. "Isotope language" of the alpine Iceman investigated with AMS and MS. *Nuclear Instruments & Methods in Physics Research, Section B: Beam Interactions with Materials and Atoms* 204:705–719.

Kwa, Chunglin. 2010. Agency and space in Darwin's concept of variation. In *Chrono-Topologies: Hybrid Spatialities and Multiple Temporalities,* ed. Leslie Kavanaugh, 79–90. Amsterdam: Rodopi.

Latour, Bruno. 1993. *We Have Never Been Modern.* Cambridge, MA: Harvard University Press.

Latour, Bruno. 1997. Trains of thought: Piaget, formalism, and the fifth dimension. *Common Ground* 6 (3): 170–191.

Lorenzi, Rossella. 2007. Blow to head, not arrow, killed Ötzi the Iceman. *Discovery News ABC.* August 31. http://www.abc.net.au/science/news/stories/2007/2020609. htm?health (accessed September 30, 2014).

Lynch, Kathleen. 2008. Archaeologies of English Renaissance literature. *Shakespeare Quarterly* 59 (2): 224–227.

MacIntyre, Ben. 2003. We know Ötzi had fleas, his last supper was steak ... and he died 5,300 years ago. *Times Online*. November 1.

Massey, Doreen B. 2005. *For Space*. London: Sage.

McGrath, Matt. 2013. World's oldest bog body hints at violent past. *BBC News*. http://www.bbc.com/news/science-environment-24053119 (accessed September 30, 2014).

MIT Technology Review. 2010. The two-dimensional arrow of biological time. *MIT Technology Review*. http://www.technologyreview.com/view/418702/the-two -dimensional-arrow-of-biological-time/ (accessed September 30, 2014).

Mol, Annemarie. 2002. *The Body Multiple: Ontology in Medical Practice*. Durham, NC: Duke University Press.

Müller, Wolfgang, Henry Fricke, Alex N. Halliday, Malcolm T. McCulloch, and Jo-Anne Wartho. 2003. Origin and migration of the alpine Iceman. *Science* 302 (5646): 862–866.

Oeggl, Klaus, Werner Kofler, Alexandra Schmidl, James H. Dickson, Eduard Egarter-Vigl, and Othmar Gaber. 2007. The reconstruction of the last itinerary of "Ötzi," the Neolithic Iceman, by pollen analyses from sequentially sampled gut extracts. *Quaternary Science Reviews* 26 (7–8): 853–861.

Oeggl, Klaus, James H. Dickson, and Sigmar Bortenschalger. 2000. Epilogue: The search for explanations and future developments. In *The Man in the Ice*, vol. 4: *The Iceman and His Natural Environment: Palaeobotanical Results*, ed. Sigmar Bortenschlager and Klaus Oeggl, 163–166. New York: Springer.

Oyama, Susan, Paul E. Griffiths, and Russel D. Gray, eds. 2001. *Cycles of Contingency: Developmental Systems Theory and Evolution*. Cambridge, MA: MIT Press.

Pappas, S. 2012. "Iceman" mummy holds world's oldest blood cells. *LiveScience*. http://www.livescience.com/20030-ice-mummy-oldest-blood-cells.html (accessed September 30, 2014).

Pfeifer, K., and Klaus Oeggl. 2000. Analysis of the bast used by the Iceman as binding material. In *The Man in the Ice*, vol 4: *The Iceman and His Natural Environment: Palaeobotanical Results*, ed. Sigmar Bortenschlager and Klaus Oeggl, 69–76. New York: Springer.

Plicht, J. van der, W. A. B. van der Sanden, A. T. Aerts, and H. J. Streurman. 2004. Dating bog bodies by means of 14C-AMS. *Journal of Archaeological Science* 31 (4): 471–491.

Ruff, Christopher B., Brigitte M. Holt, Vladimir Sládek, Margit Berner, William A. Murphy Jr., Dieter zur Nedden, Horst Seidler, and Wolfgang Recheis. 2006. Body size, body proportions, and mobility in the Tyrolean "Iceman." *Journal of Human Evolution* 51 (1): 91–101.

Science Daily. 2011. A rest, a meal, then death for 5,000-year-old glacier mummy: Scientists consolidate results of research into Ötzi's state of health and his death. http://www.sciencedaily.com/releases/2011/10/111025091533.htm (accessed September 30, 2014).

Seiler, Roger, Andrew I. Spielman, Albert Zink, and Frank Rühli. 2013. Oral pathologies of the Neolithic Iceman c. 3,300 BC. *European Journal of Oral Sciences* 121 (3.1): 137–141.

Sjovold, Torstein, Wolfram Brenhard, Othmar Gaber, Karl-Heinz Kunzel, Wernre Platzer, and Hans Unterdorfer. 1995. Verteilung unf Grose der Tatowierungen am Eisman vom Haulabjoch. In *The Man in the Ice*, vol. 2, ed. K. Spindler, E. Rastbichler, H. Wilfing, D. z. Nedden, and H. Nothdurfer, 269–273. Vienna: Springer.

Skeates, Robin. 2013. Communication over space and time in the world of the Iceman. In *Space and Time in Mediterranean Prehistory*, ed. Stella Souvatzi and Athena Hadji, 138–159. London: Routledge.

Smolin, Lee. 2013. *Time Reborn: From the Crisis of Physics to the Future of the Universe*. London: Penguin Books.

South Tyrol Museum of Archaeology. 2013. Anatomic anomalies. *South Tyrol Museum of Archaeology*. http://www.iceman.it/en/node/259 (accessed September 30, 2014).

Spindler, Konrad. 1994. *The Man in the Ice: The Preserved Body of a Neolithic Man Reveals Secrets of the Stone Age*. London: Phoenix Books.

Star, Susan Leigh. 1990. The sociology of the invisible: The primacy of work in the writings of Anselm Strauss. In *Social Organisation and Social Processes: Essays in Honour of Anselm Strauss*, ed. David R. Maines. New York: Aldine.

Stromberg, Joseph. 2013. What is the Anthropocene and are we in it? *Smithsonian Magazine*. January.

Thompson, Evan. 2007. *Mind in Life: Biology, Phenomenology, and the Science of the Mind*. Cambridge, MA: Belknap Press of Harvard University Press.

Travis, John. 2011. How does the cell position its proteins? *Science* 334 (6059): 1048–1049.

Turnbull, David. 2007. Maps, narratives, and trails: Performativity, hodology, distributed knowledge in complex adaptive systems—an approach to emergent mapping. *Geographical Research* 45 (2): 140–149.

Turnbull, David. 2008. String and stories. In *Encyclopaedia of the History of Science, Technology, and Medicine in Non-Western Cultures*, ed. Helaine Selin. Berlin: Springer.

Vanzetti, A., M. Vidale, M. Gallinaro, D. W. Frayer, and L. Bondioli. 2010. The Iceman as a burial. *Antiquity* 84 (325): 681–692.

von Baeyer, Hans Christian. 2004. *Information: The New Language of Science.* Cambridge, MA: Harvard University Press.

Wolfe, Nathan. 2012. What's left to explore? *TED.com.* http://www.ted.com/talks/nathan_wolfe_what_s_left_to_explore?language=en (accessed September 30, 2014).

Wrangham, Richard. 2009. *Catching Fire: How Cooking Made Us Human.* New York: Basic Books.

Zink, Albert, Angela Graefen, Klaus Oeggl, James Dickson, Walter Leitner, Günther Kaufmann, Angelika Fleckinger, Paul Gostner, and Eduard Egarter-Vigl. 2011. The Iceman is not a burial: Reply to Vanzetti et al. (2010). *Antiquity* 85:328.

9 Beyond the Life/Not-Life Binary: A Feminist-Indigenous Reading of Cryopreservation, Interspecies Thinking, and the New Materialisms

Kim TallBear

Cryopreservation—or the deep freezing of tissues—enables storage and maintenance of biospecimens from whole human bodies, plant materials, and blood samples. It allows the suspended animation and temporal transport of cells, and within them DNA, into realms beyond the bodies whose lives these biologicals once helped constitute. That we can barely read that code matters less, as Joanna Radin has pointed out, than scientists' desperate desires "to accumulate fragments of a world whose inherent plasticity, augmented by the corrosive forces of modernity, seemed poised to render certain life forms extinct" (Radin 2012). Taking blood is sometimes seen as urgent—not to derive data immediately, but to preserve it in the face of the expected extinction of peoples for future yet unarticulated research questions and for future technologies of analysis. Cryopreservation's co-constitutive narrative with indigenous genetics has long been that gathering indigenous biomaterials is about staving off certain death. The narrative calls for preserving remnants of human groups and their nonhuman relations, defined in molecular terms, and archiving those molecular patterns and instructions before peoples or species "vanish" in death by admixture, or actual extinction in the case of nonhuman species.

Research ethics are changing of late. Geneticists increasingly do collaborative research with indigenous peoples and attempt to account for indigenous jurisdiction. Their collaborations involve figuring out how to ethically reuse old tissues for new purposes, including techniques of reconsent (Kowal 2013; Radin 2017). I am encouraged by such interventions. However, structural inequities between indigenous peoples and those who would study and manage their bodies, lives, and biologicals are entangled with problematic dominant narratives that continue to condition even newer collaborative research. The supposed vanishing of "traditional" or indigenous peoples in the face of globalization remains a common narrative in genetic science. Fundamental to the inquiry I make in this chapter is to examine the role of death in de-animating living indigenous peoples. The concept of death, ironically, is integral to the constitution of life as molecular in both scientific and popular discourses. The molecular definition of life is built upon indigenous bodies that, because they are expected to vanish, do not have to be considered as among the living beneficiaries in the promises of genomic futures (Reardon and TallBear 2012).

This chapter offers three critiques of cryopolitics that highlight how long-standing fundamental assumptions and structural inequities continue to shape the sampling and study of indigenous bodies and the biologicals derived from them. Specifically, nonindigenous binary concepts of life versus death and human versus nonhuman continue to shape not only assumptions about indigenous bodies but also relations between scientists and indigenous peoples, even in an age of more just ethical practices. The chapter links the cryopolitical project of maintaining genetic material from indigenous bodies with other academic conversations—"animal studies" and the "new materialisms"—that, to different degrees, labor under and seek to repair the same life versus death and animate versus de-animate dichotomies that plague genetic research on indigenous peoples.

In general usage, the terms *animate* and *inanimate* reflect a categorical divide between entities—those that are seen to live versus those that are deemed to be not alive, and this is in the West defined at the level of the organism. I use the term *de*-animate after Mel Chen's use in *Animacies* (2012). Chen derives the concept of animacy from linguistics. A "hierarchy of animacies" refers to the greater and lesser relative degree of entities' sentience, aliveness, (self-)awareness, and agency. The animacy hierarchy is actualized through the associated verbs/adjectives "animate" and "de-animate" that refer to the greater and lesser aliveness attributed—in the nonindigenous knowledges I interrogate—to some humans over others, and to humans over nonhumans.

Indigenous standpoints confound the Western animacy hierarchy. The academic conversations highlighted in this chapter seek new language to articulate relations between humans and nonhumans, and I argue that they will benefit from indigenous standpoints that never forgot the interrelatedness of all things. The chapter ends with an example of the relations between certain indigenous people and a vibrant material, pipestone, in ways that challenge hierarchies of life. This chapter constitutes an intervention in Western academic thought. However, my ultimate goal is to support the flourishing of indigenous thought and life in the twenty-first century. The academy is integral to the colonial state that manages indigenous lives and nonhuman relations, too often to our collective detriment.

First Critique: Cryopreservation Is Implicated in the Colonial Appropriation of Indigenous Natural Resources

As Joanna Radin has explained, the International Biological Program (IBP) that ran from the early 1960s through 1974 set the stage for the biological sampling of indigenous peoples and nonhumans as remnants of a bygone world threatened by an increasingly "civilizing" yet toxic world. IBP scientists laid out a goal to research "primitive isolates ... to salvage information that might benefit civilized communities' understandings of themselves" (Radin 2013, 496). But that conception of indigenous bodies as "natural resources," the raw materials upon which nations are built, did not begin in the mid-twentieth century. Even prior to US Independence, the nation-building project relied on the appropriation of indigenous peoples' lands. In the eighteenth and nineteenth centuries, the United States positioned itself—positioned "Americans" or whites—as the rational agents capable of transforming nature into productive property, and indigenous peoples as incapable of developing, indeed even surviving, in the face of the modern industrial state (Harris 1993). Colonial spokespersons—both government agents and anthropological and biological scientists—have continued to rehearse that death song for a long time.

In the twenty-first century, building on the molecular project begun with the IBP, the goal is to transform natural resources that are genomes into something of value for "all humanity," genomic knowledge "for the good of all" (Reardon and TallBear 2012). That is the mantra. But who, in pragmatic terms, counts as human? In other words, who gets to figure as a potential recipient of those benefits? Certainly a vanishing, let alone a vanished, indigene cannot make a claim to future benefits, and is not included among the modern humanity that is the "we" addressed by genomic storytellers. As Jenny Reardon and I have argued at length elsewhere, twenty-first-century human genome research promotes itself as antiracist because it undermines biological conceptions of race. Yet, paradoxically, it relies on making moral and ultimately racially hierarchical claims to the natural resources of indigenous peoples. Witness the as-always emblematic language and imagery of National Geographic's Genographic Project, when population geneticist Spencer Wells addresses Australian Aboriginal painter Greg Singh:

What I'd like you to think about with the DNA stories we're telling is that they are that. They are DNA stories. It's our version as Europeans of how the world was populated, and where we all trace back to. That's our songline. We use science to tell us about that because we don't have the sense of direct continuity. Our ancestors didn't pass down the stories. We've lost them, and we have to go out and find them. We use science, which is a European way of looking at the world to do that. You guys don't need that. (Tigress Productions 2003)

In the film, Wells, the Genographic Project principal investigator, responds to Singh's dubiousness that his Aboriginal ancestors trace back to Africa. Singh insists on the veracity of Aboriginal origins on the continent of Australia: "We know our stories. We know about creation. We know we come from here" (Tigress Productions 2003). Conversely, US American Wells explains that he and his "European" people are still searching for their origins. You have your stories, he says to Singh. We just want ours. In an interesting reversal of the usual narrative, "Europeans" are presented as disadvantaged in relation to indigenous people. It is indigenous people who potentially "take away" if they deny their DNA, a resource without which Wells and his people will lose their past. While the narrative has an interesting new twist, the old familiar pattern remains intact, whereby a "European" makes a moral claim on indigenous natural resources—this time, indigenous DNA.

Second Critique: Cryopreservation Aims to Preserve Indigenous DNA, but Is Predicated on Indigenous Death

Human genome diversity research deploys the concept of indigenous peoples' perpetually impending death in order to support a genomic rearticulation of indigenous life as the rightful patrimony of global society. Indeed, indigenous vanishing and dying fosters emerging genomic versions of life. And because indigenous peoples are not expected to survive—this is part of the "logic of elimination" referred to by Patrick Wolfe in his analysis of settler-colonial societies and their appropriation of indigenous land and resources—they do not have to be factored into the promises of genomic futures (Wolfe 2006).

Within that, cryopreservation of indigenous biological samples is about the extension, study and preservation of life. But it is a notion of life—*molecular* life—focused on largely indecipherable patterns and instructions that comprise DNA's scaffolding. It is a materialist conception of life that I do not disagree with. It even has bibliographic beauty, as high-tech visualizing technologies reveal great dynamic and productive molecular archives. But cryopreservation also amplifies the poverty of the genomic articulation of indigeneity, which becomes ever more salient in our world of genomics-as-nation-building (TallBear 2013a). Cryopreservation, like genomic indigeneity, emerges not only out of technoscientific innovations but also from a discourse of endangerment and death—a long-standing narrative of indigenous extinction that has pervaded Western culture for several centuries (Berkhofer 1986; Dippie 1991; Human Genome Diversity Workshop 1 1992; Human Genome Diversity Workshop 2 1992; Genographic Project n.d.).

Michel Foucault's notion of biopower helps elucidate the role of cryopreservation in the work of making indigenous DNA live while simultaneously narrating indigenous peoples' death. Biopower, in Foucault's telling, emerged at the end of the eighteenth century and took hold by the nineteenth, replacing the "ancient right [of the sovereign] to *take* life or *let* live" with "a power to foster life or disallow it to the point of death" (Foucault 1990, 138).

Consequently, biopower focuses on understanding morbidity, not the old spectacle of death under the sovereign, but rather illnesses and "permanent factors" that perpetually weaken life (Foucault 2003, 243–244). Biopower produces statistics (e.g., fertility, disease and morbidity rates; race and ethnicity data) about the biological "populations" it seeks to regulate, and these statistics help to bring them under control. "Population" is a key concept that conditions genomic sampling and the (re)articulation of indigenous peoples according to genomic knowledge. Indigenous peoples are, for all practical purposes, conflated with the concept of a biological population in the course of research. I'll come back to this problem shortly.

The emergent multicultural global genomic subject is a product of two apparently contradictory beliefs: the pervasive and enduring sense of the inevitable, impending death of the Native American and other indigenous peoples in the face of Western "civilization" and the belief that continued indigenous existence is crucial for US nationalist coherence. Indigenous DNA, as a proxy for bodies and for life itself, helps bring death within the realm of power. In these interwoven narratives, indigenous death is far enough along to justify appropriating indigenous resources but is ultimately held back from the brink of extinction when indigenous genome knowledge is produced for consumption by a twenty-first-century knowledge society (Reardon and TallBear 2012). The promise of human genome diversity is, in a very real sense, the promise of eternal life.

Indigenous DNA, in and of itself, when found in an individual's genome, can be used to signify the life and essence of long-dead, often nameless indigenous ancestors. Or genomic knowledge more broadly constituted in part through examining indigenous DNA in relation to that of diverse populations around the world can help connect living people to founding populations—to ancestors. As Kowal (2013) describes, this can lead to an intensification of the vitality of genome knowledge consumers; part of the appeal of learning one's "deep ancestry" is the acquisition of new kin relations. In short, the death of indigenous people(s) is simultaneously always occurring but ultimately avoided as genome knowledge seekers tap into indigenous lives signified in DNA that has been found to be dispersed among individuals who did not previously identify as indigenous. Thus can the conscience be soothed of a global society built in no small part on the extraction of indigenous resources.

Cryopreservation and its disappearing indigene narrative aid a broader genomic death song sung ironically while indigenous peoples circulate more actively in genome worlds and other global networks. With new ways of caretaking biological samples—for example, DNA on loan with the property interests staying with donors, rigorous efforts to reconsent old samples, and indigenous-managed biobanks—we see a partial disruption of hierarchies between scientists and their indigenous subjects and a complication of the indigenous death narrative (Arbour and Cook 2006). Emma Kowal (2013) points out that biological substances derived from indigenous bodies are generative of social relations between scientists and indigenous peoples, both of whom have interests in and desire some control over those substances.

Indigenous people are "animated" when they become actively engaged in the production of genomic knowledge, either as scientists or as collaborators, and when regulators pay new attention to building networks or to training indigenous people to do the science. Yet, this animation is mitigated owing to the fact that biovalue, or the production of scientific value from human body parts, is "surplus value" actually produced when "marginal forms of vitality ... [including] the bodies and body parts of the socially marginalized—are transformed into technologies to aid in the intensification of vitality for other living beings," those so-called modern humans who seek to benefit from genome knowledge, both scientists and the consumers of their knowledge products (Waldby 2000 cited in Kowal 2013, 581). In this way, indigenous bodies are de-animated in the hierarchies of "civilizing" scientific knowledge production while simultaneously reanimated when living indigenous people are encouraged to become scientists either in actual practice or primarily in name when they are cultivated as coauthors or co-principal investigators within new, more collaborative ethical regimes (Kowal 2013, 591).

Third Critique: Genomic Indigeneity Is an Inadequate Form of Indigenous Resistance to the Death Narrative

Genomic indigeneity is produced from the analysis of preserved samples extracted from populations defined by scientists as of particular value owing to their geographic and cultural isolation, and in particular, their longevity in particular environments. The population geneticists who developed this mode of knowledge production often regarded indigenous people as frozen in time or relics of the past (Radin 2013). Scientists' valorization of deep connections to place overlap partially with indigenous groups' own articulations of peoplehood, although indigenous people's co-constitutions with place are not fully accounted for in genome science accounts. Indigenous peoples do emphasize their long-standing relations to place, but in ways that differ significantly from many genome scientists. What differentiates the accounts of indigenous peoples is that they emphasize complex forms of relatedness of peoples and nonhumans in particular places. We might say, in the parlance of science and technology studies (STS), that indigenous peoples consider identity to be the product of a co-constitution of human and nonhuman communities while human genome diversity focuses predominantly on the movements of human populations through landscapes to populate the world (Reardon 2001; Jasanoff 2004; Latour 1999).

To be sure, the impact of environmental factors on the selection and mutation of genetic markers figure in decisive ways in genomics. However, as human genomic markers made to stand for "populations" are narrated as moving along ancient continental "migratory" routes, human agency upon the land and linear migratory narratives through time and space are still foregrounded over deep human–nonhuman relations *in place*. Contemporary "admixed populations," in such a figuration, come to be conceived of as the descendants of indigenous "founder populations." The primary agents in such narratives are abstracted human bodies and populations that are retrospectively modeled as biologically reproducing themselves along a linear path through time and space.

The dismissal by nonindigenous thinkers of indigenous origin stories in place, I argue, is not simply the result of privileging scientific evidence over indigenous creationism and rejecting "religious" accounts broadly. Such dismissals also ignore the importance of indigenous peoples' emphasis on land-human co-constitutive relations. For example, when indigenous peoples push back against scientific narratives of indigenous Americans' genetic "origins" in "Africa" or "Siberia," indigenous peoples are not simply being antiscience. They may, in fact, be very interested in the science. At stake in contesting such genomic narratives is the desire of indigenous peoples to emphasize their emergence as particular cultural and language groups *in social and cultural relation* with nonhumans of all kinds—land formations, nonhuman animals, plants, and the elements in very particular places—their "homelands" or "traditional territories," for example.

Furthermore, genomic indigeneity also fundamentally contradicts a definition of "indigenous" in its explicitly political formation. Indigenous peoples, in their networking between groups, have demonstrated that "indigeneity" can serve as an organizing category *added to* group-specific understandings. The concept "indigenous," forged by delegates from communities around the world in transnational institutions like The Hague or the United Nations, helps individual communities articulate our collective resistance to the assimilative tendencies of the nation-state (Niezen 2003). The term "indigenous," in this sense, is about survival, a "we are here and proliferating" discourse. Native American studies scholar Elizabeth

Cook-Lynn defines indigeneity as not simply "a political system based in economics and the hope for a fair playing field. Nor is it a belief system like religion. It is, rather, a category of being and origin and geography, useful for refuting other theories of being and origin" (e.g., those of Christianity and of science). "Today," she argues, "indigeneity may be thought of as the strongest focus for resistance to imperial control in colonial societies." Furthermore, she concludes, indigenous peoples, as a class, are "expanding rather than vanishing or diminishing" (Cook-Lynn 2012, 15).

Indigenizing Interspecies Thinking

Thus far, I have focused on how a cryopolitical framework can be used to make sense of genome scientists' engagement with indigeneity. I turn now to discuss how indigenous standpoints can be used, more broadly, to examine how questions of animacy are foreclosed by cryopolitical impulses in other fields of knowledge. Indigenous standpoints accord greater animacy to nonhumans, including nonorganisms, such as stones and places, which help form (indigenous) peoples as humans constituted in much more complex ways than in simply human biological terms. Indigenous peoples are not vanishing, but are vibrant and growing in number and power. Nor are we are oppositional, biological, or cultural relics of an ancient and dying past, now absorbable into Western bodies, institutions, and definitions.

Recently there has been an upsurge of scholarship in the field of "animal studies," in which social science and humanities scholars attempt to recover knowledge territory claimed for several centuries by the natural sciences, and to lessen hierarchies between "Westerners" and their nonhuman others. A good example is *multispecies* ethnography in which scholars apply anthropological approaches to studying social relations between humans and nonhumans. In 2010, Eben Kirksey and Stefan Helmreich wrote in a *Cultural Anthropology* special issue on the subject, that new anthropological accounts are starting to appear in which nonhumans previously relegated to the status of "bare life" or "that which is killable" are now considered "alongside humans in the realm of *bios*, with legibly biographical and political lives." Multispecies ethnography shows how "'organisms' [e.g., animals, plants, fungi, microbes] shape and are shaped by political, economic, and cultural forces" (Kirksey and Helmreich 2010, 545).

New work in critical animal studies articulates with Dorion Sagan's work on "interspecies communities" and his critique of overly linear evolutionary narratives. Sagan, a well-known science writer and son of astronomer Carl Sagan and biologist Lynn Margulis, seeks to dismantle hierarchies in the relationships of Westerners to their nonhuman others. Speaking of symbiogenesis and the importance of microbes in constituting humans, Sagan explains:

> We are crisscrossed and cohabited by stranger beings, intimate visitors who affect our behaviour, appreciate our warmth, and are in no rush to leave. Like all visible life forms, we [humans] are composites. (Sagan 2011, 6)

This account of symbiogenesis seems to lean toward an ethos of "we are all related." I read in Sagan that we are all of us—human and nonhuman—networked sets of *social-biological* relations. He calls such human–nonhuman networks "interspecies communities," which resonates with what Vine Deloria Jr. called an "American Indian metaphysic," a "set of first principles we must possess in order to make sense of the world in which we live" (Deloria 2001). In his essay of the same title Deloria wrote that the best description of this term is

> the realization that the world, and all its possible experiences, constituted a social reality, a fabric of life in which everything had the possibility of intimate knowing relationships because, ultimately everything was related.

Yet Deloria's and Sagan's understandings of intimate human–nonhuman relations are not in complete harmony. In the opening of his 2010 American Anthropological Association keynote, "The Human Is More Than Human," Sagan turned his attention to "life." He opened:

> Well it is to this universe that I want to turn again, and to a specific part of it. I want to turn to life, and within that part a fascinating subsystem, the one in which, of course, we are most interested. That is, humanity, ourselves. (Sagan 2011)

For Sagan, life is limited to things that are organismically defined. This tends to be true as well of social scientists and humanists who engage in animal studies, and not only biophysical scientists. Multi- and interspecies conversations restrict attention to beings that "live" according to a definition that cannot completely contain indigenous standpoints. In an "American Indian Metaphysic," as Deloria describes it, the notion of "interspecies" must be expanded to include nonhuman others that are not understood in critical Western frameworks as living.

An example lies in the knowledge production of my colleague Craig Howe. Howe is an Oglala Lakota architect, anthropologist, and Lakota studies expert who left the mainstream academy to found the Center for American Indian Research and Native Studies (CAIRNS), which he built on his family's land between the Rosebud and Pine Ridge Indian reservations in South Dakota. In addition to running Lakota and Native American studies workshops at the facility, Craig lives there in daily contact with a host of nonhumans on the land, and he documents those relations weekly in his blog, *Oko Iyawapi* ("week count").[1]

There, Howe engages with rabbits who pointedly observe him when he works outside, who alternately take afternoon naps in the shade of his structures, and sometimes thwart his architectural projects with their own work—their adamant digging. He studies the lodges of spiders, akin to tipis in their form. He follows the movements of star people in relation to distant buttes as he plots and builds the structures of the CAIRNS facility. One could see Howe as "studying" rabbits, spiders, and the cosmos—entities less animate than his human self. But such hierarchy is not found in Howe's words and photographs. Rather, in his blog, Howe documents his relations with nonhumans, including some currently excluded in animal studies, and the ways in which they do work in the world and help constitute the land and the buildings he inhabits along with them.

What Is not New about the New Materialisms? An Indigenous Metaphysic: Networked Sets of Social-Material Relations

There is a second academic conversation in which nonindigenous thinkers address the divide between life and not-life, and expand conceptions of humans' intimate relations. Work in the new materialisms seeks to extend our understanding of how nonhuman lives press into and are co-constituted with human lives and are therefore agentic in our world. It stands in the space where animal studies and multi- and interspecies thinking fall short. In *New Materialisms: Ontology, Agency, and Politics* (2010), Coole and Frost explain the project:

In terms of theory itself, finally we are summoning a new materialism in response to a sense that the radicalism of the dominant discourses which have flourished under the cultural turn is now more or less exhausted. We share the feeling current among many researchers that the dominant constructivist orientation to social analysis is

inadequate for thinking about matter, materiality, and politics in ways that do justice to the contemporary context of biopolitics and global political economy. (Coole
and Frost 2010, 6)

Similarly, Jane Bennett in *Vibrant Matter* calls for us to take seriously the
vitality of things, meaning "the capacity of things—edibles, commodities,
storms, metals—not only to impede or block the will and designs of
humans but also to act as quasi agents or forces with trajectories, propensi-
ties, or tendencies of their own" (Bennett 2010a, viii). She also refers to
this as an "enchanted materialism" (Bennett 2010b, 9). Why should we
care about this? Like Coole and Frost, Bennett suggests that taking seriously the vitality of nonhuman, *nonliving* things might shift our political
responses to public problems. In this discourse, we find echoes of how
genome scientists have been pushed to see the vitality of indigenous peoples—that they/we are more than de-animated resources for scientific and
national vitality. Bennett and other humanists as well as social scientists
are reappraising the importance of materiality to human social and cultural life. The world is mediated heavily through culture—increasingly so
in the Anthropocene—but the materiality of that world is also productive,
not simply static, dead, and acted upon. The new materialists hold that, to
really grasp the nature of and potential solutions for the world's most critical problems, including environmental degradation, climate change, poverty, systemic violence, and warfare, nonhumans in all their myriad forms
must be given their due.

In *Animacies*, Mel Chen has a similar intellectual-political agenda. After
an extended illness forced a rethinking of the boundary between life and
death, Chen interrogated that "fragile division between animate and inanimate." In *Animacies*, Chen disrupted the stubborn binaries of life/death,
human/animal, dynamism/stasis, subject/object that underlie violent hierarchies in our world to invoke the language of *animacy* in place of *life* or
liveliness. While I am more comfortable using the term "life" in ways that
go beyond biological meaning as part of my understanding of animacy, for
Chen, who would understand "life" probably more strictly in biological
terms, animacy is defined in relation to "agency, awareness, mobility, and
liveness" (Chen 2012, 1–2). "Animacy is much more than the state of being
animate" (4). Like Bennett, Chen writes of the animacy of metal, more
specifically mercury and lead, that can poison human and nonhuman
bodies.

Mel Chen and I have discussed our similar desire to disrupt the human/ animal, life/not-life divides. Among the new materialists, Chen thinks more than others about the role of indigenous thought in understanding and disrupting "animacy hierarchies" (Chen 2012, 24–28). This is unsurprising because Chen's particular project is to explore the roles of race and sexuality in shaping "animacy as a specific kind of affective and material construct that is ... nonneutral in relation to animals, humans, and living and dead things" (5). Yet Chen also seeks a "secular" (Chen's word) language to express the animacy of nonhumans beyond dumb or lifeless materiality, as does Bennett, who explains that "material vibrancy is not a spiritual supplement of 'life force' added to the matter said to house it" (Bennett 2010a, xiii). On the other hand, as an indigenous person, I am comfortable enfolding spirits or souls into descriptions of the beingness of nonhumans. As Charles Eastman wrote in his 1911 book *The Soul of the Indian*:

The elements and majestic forces in nature, Lightning, Wind, Water, Fire, and Frost, were regarded with awe as spiritual powers, but always secondary and intermediate in character. We believed that the spirit pervades all creation and that every creature possesses a soul in some degree, though not necessarily a soul conscious of itself. The tree, the waterfall, the grizzly bear, each is an embodied Force, and as such an object of reverence. (Eastman [1911] 1980, 14–15)

Eastman was born into an "Indian" world, but at a moment when Indian people were dispossessed from their traditional lands in what is today Minnesota. Some were killed. Others were imprisoned after an 1862 war against settlers, and some fled to Canada. Eventually many ended up on reservations in South Dakota and Minnesota. Eastman converted to Christianity and trained in Boston as a medical doctor. In the early twentieth century, he wrote of American Indian life and history from an Indian standpoint (Eastman 1915). In the passage above, he explains an American Indian "spiritual" approach to the US Christian reader.

Both Eastman and Deloria, then, can be understood within a framework that posits social relations not only between humans and "animals," but also between humans and "energy," "spirits," "rocks," and "stars," in the constitution of American Indian knowledge about the world. Deloria called this an "American Indian metaphysic." I'll call it an indigenous metaphysic: an understanding of the intimate knowing relatedness of all things (Deloria 2001). As I refer to them, the co-constitutive entanglements between the material and the immaterial—that is, indigenous peoples' social relations also with "spirit" beings (for lack of a better term) described by Eastman and Deloria—constitute a boundary crossing that is difficult for interspecies and new materialist thinkers. Nonindigenous society has put much effort into erecting a barrier between what it is thought humans can *know* through their materialistic, empirical investigations and what (some) humans *believe* to exist beyond the knowable material world. This knowing/belief divide, as Paul Nadasdy points out, is a form of discrediting language used, for example, by even sympathetic anthropologists when explaining indigenous subjects' cosmologies (Nadasdy 2003). The same derisive binary of "knowledge" (often claimed to be for the good of all) versus the "beliefs" of indigenous peoples is commonly cited when indigenous people protest, for example, genome research on their blood or on the bones of their ancestors. It is a binary that certain genome scientists, but also members of the public, use in turn to express disdain for indigenous resistance to particular forms of research.

Eastman and Deloria get classified as American Indian intellectuals—the pertinent racial category of the twentieth century—but they were both in fact Dakota. The "American Indian" pieces they wrote came out of a disproportionately Dakota cultural background. We Dakota are part of a broader cultural group, the Oceti Sakowin that also includes Lakota and Nakota peoples. We have been mislabeled by Europeans, Americans, and others as "Sioux." Eastman and Deloria's work is useful for thinking about these things and is already widely known. It would be exciting to see thinkers whose work concerns the intellectual production of other indigenous peoples add to this conversation. Take, for example, the recent work of another scholar of (Yoeme) indigenous thought, David Delgado Shorter. He provides us language that goes beyond the "spiritual" for capturing the kinds of "interspecies" (perhaps "cross-beings" is a more inclusive term) forms of relating that indigenous peoples have engaged in:

Critical analyses of the oral stories and earliest ethnographic accounts continue to reveal that rather than being "spiritual," native people were relating. … Less cumbersome than "intersubjective," "related" might best replace "spiritual" since it neither denies possible theisms or hierarchies. One's relations are not solely related by blood, or tribal heritage. They are the families chosen and not chosen. They are the humans and other-than-humans sharing and withholding power. They counter solitude with solidarity. They provide meaning and identity beyond the confines of race, tribe,

and species. In these ways, being related offers more than any abstract notion of religion or "spirituality" ever could. (Shorter 2016, 20)

Indigenous thinkers have important contributions to make to conversations in which human societies rethink the range of nonhuman beings
with whom we see ourselves in intimate relation and, precisely because of
the varied ways in which indigenous peoples relate, our possibilities for
being the world. The advantage of indigenous analytical frameworks that
are not secular is that they are more likely to have kept sight of the profound influence in the world of beings categorized by Western thinkers
(both the church and science) in hierarchical ways as animal, or less animate. Now that theorists in a range of fields are seeking to dismantle those
hierarchies, we should remember that not everyone needs to summon a
new analytical framework or needs to renew a commitment to "the vitality
of [so-called] things." Indigenous standpoints that never constructed hierarchies in quite the same way can and should be at the forefront of this new
ethnographic and theoretical work. We can converse with the existing work
and bring additional insights.

Bringing indigenous thought into these conversations does not simply
increase intellectual rigor and expand multiculturalism in the academy.
And by "indigenous thought" I do not mean some static notion of indigenous "traditional" knowledge, but rather engagement with the thinking
that *living* indigenous people do today. Infiltrating the academy is but a step
toward a more important goal, which is to construct disciplines that better
serve our interests (and the interests of others who have been historically
marginalized) (Harding 2008). The disciplines are looked to by a knowledge
society to support policy decisions—to address pressing societal problems.
Indeed, considerable damage has been done to both indigenous people and
our nonhuman relations, as we have been displaced from traditional lands;
have had our communal land-holding disrupted; been subjected to related
assimilationist policies, including forced conversions to Christianity and
the imposition of a state/religion spirit/material divide that fundamentally
misunderstands indigenous ontologies; and finally, had our traditional
lands subjected to the environmental degradation that accompanies colonial appropriation.

The Idle No More indigenous social movement founded in Canada by
four Canadian women—three of them indigenous—has spread around the
world. It foregrounds links between the colonial state's violations of

indigenous rights and extractive colonial industries' violations of the Earth (Kino-nda-iimi Collective 2014). Idle No More is a social movement that recognizes the intimacies of human and nonhuman sociality. Articulating indigenous thought within academic and policy frameworks matters for indigenous people. We cannot allow the old science/religion, and all of those other human/animal, life/not-life divides—upheld by institutions to govern our lives, the land, and the lives of nonhumans who have been savaged by Western analytical frameworks, animacy hierarchies, and the institutions they produce—to continue to pervert understanding of our lifeways.

Also important for twenty-first-century indigenous thriving is that some indigenous communities are working to find new languages and approaches to apprehending knowledge and building indigenous-controlled institutions using conceptual frameworks that combine both disciplinary and indigenous knowledges. Tribal environmental science and ethnobotany programs and tribally controlled educational institutions in the United States come to mind. As regulatory, professional, and educational practices are being developed, such programs subject to critical reexamination the normative science/religion divide that is so integral to Western academic thought (Deloria 2001). Those very words "science" and "religion," in part or in whole, are inadequate to describe the world and are inadequate for doing the knowledge work that indigenous communities do. Within the context of such programs, for example, what insights and approaches are developed when tribal environmental science programs do "ethnobotany" along with valuing traditional knowledge-getting methods? For example, what insights are added by being receptive to or seeking knowledge about the nonhuman world brought to us by spirits—relations that science may never see or measure—in dreams or in ceremonies?

Life, Blood, and a Sacred Stone

I will now turn from nonhuman organisms, from multispecies relations, and from indigenous DNA to pipestone. I spent the last dozen years studying the life and death—both material and symbolic—that inhere in the production of Native American DNA from blood samples frozen out of place and time. Both blood and the DNA it carries are material-semiotic, constituted of molecules and signs—metaphors and long-standing narratives of indigenous vanishing—that make "indigenous blood" and "our DNA" coherent concepts and things of value. It may surprise readers that in beginning ethnographic work at the pipestone quarries in the rural southwest of the State of Minnesota in the north central United States, I am moving into conceptual territory that is not so far from DNA.

Starting with DNA and cryopreservation helps me tell the intellectual history of how I got here. I want to investigate the life that inheres in a particular stone, and the social relations that proliferate as that stone emerges from the earth, is carved into pipe, and is passed from hand to hand. Readers may have heard them called "peace pipes." That is not what we call them. Pipestone is otherwise known as Catlinite after the nineteenth-century US American artist George Catlin, who painted the site in southwest Minnesota where the red stone is cut from the earth. There is a story that tells us about the blood in that stone, a story that along with our loving attention to its materiality enables us to apprehend pipestone's vibrancy, its fundamental role in our peoplehood. Indeed, the pipestone quarries comprise a site that is central to Dakota peoplehood. Like bioscientists in the twentieth and twenty-first centuries with their imperative to bleed indigenous peoples before it was too late, a nineteenth-century euro-American painter and early twentieth-century geologists and government agents saw the place where the red stone lies as an artifact of a waning culture and time. They produced a "National Monument" to conserve it. The binaries at play then and today are past/present, traditional/modernity, alive/dead and dying.

The pipestone is a vibrant material. But in contrast with indigenous DNA, I am not referring here to cellular vibrancy. We can describe pipestone as vibrant because without it prayers would be grounded, human social relations impaired, and everyday lives of quarriers and carvers depleted of the meaning they derive from working with stone. That story is a flood story in which a young woman is the only one to survive atop a hill, and the blood of all of her dead people—once the waters recede—are pooled in one place that becomes the *cannupa ok'e*, the blood of the people or the red stone.[2] And so this place is taken by many to be sacred. The *cannupa ok'e*, the quarry, is special not only to Dakota people, but to other indigenous people who dig there today, and to others for whom the pipe is also central. The stone there is sometimes spoken of as a relative, harkening in part back to that creation narrative.

US Park Service pamphlets from the Pipestone, Minnesota quarry represent pipes as artifacts, as craft objects, and detail the geology and history of white incursion in the quarry in the late nineteenth and early twentieth centuries, as well as the regulatory response of the US government. These material and regulatory histories are important, but as with the production of indigenous biological samples abstracted from living bodies and vibrant peoples, Western politics and knowledges surrounding the quarry in effect maintain a knowledge binary between the material and immaterial. That said, in good liberal multiculturalist tradition, the Park Service also acknowledges indigenous spiritual beliefs and practices related to pipestone.

Indigenous people at the quarries resist binary narratives, to a degree. Dakota quarriers/carvers reject narratives of dead and vanished Indians, past versus present. However, the "sacred" and, related to this, the "traditional" are key concepts for some of those quarriers and for many of our people. The stone is also sometimes spoken of as *a relative*. The quarries are at the centers of quarrier lives *regardless* of whether they are referred to as sacred or not. However, the idea of the sacred is a big fish in the rushing waters surrounding Pipestone Quarry discourse. It is not only science that relies on a material/immaterial binary; the discourse of the "sacred," taken up by indigenous peoples ourselves, also furthers such thinking.

The Park Service and some concerned federally recognized tribes currently struggle with ethical issues raised by the commercial sale of pipestone objects at the quarries. There is a debate taking place in Indian Country and in federal meetings about "sacred" stone that can only be seen as such if we indigenous peoples simultaneously embrace the idea of the profane. This stance is potentially at odds with a view of relationality that would acknowledge that indigenous peoples and the stone have long existed in more intimate and complex sets of relations than the notions of "sacred" and "profane" can represent.

To loop back to the topic of DNA, just as we indigenous peoples insist on our continuing survival, in part, through involvement in managing our own DNA, indigenous quarriers and carvers, medicine people, and everyday people who pray insist on living with the red stone daily. And they make decisions—some of them seen as compromises about how to best work with the vibrant objects of their attention. For example, while some indigenous people agree to engage in research or commercial activities related to DNA, others sell pipestone jewelry and craft pieces to make a living. It is a complex world in which indigenous peoples sometimes make decisions to use things considered sacred in some instances (DNA or pipe-stone) in nonsacred ways that enable their survival in other ways. Indeed, describing objects and practices as sacred may also be a strategy for survival or a particular historically conditioned form of remembering that may not align precisely with our ancestors' precolonial understandings of how to regard an object or practice. Literal red human blood (that holds indige-nous peoples' DNA so important for research) and the blood-red stone have been translated into "resources." While this characterization is troubled by Dakota and other indigenous quarries and carvers, it is not undone.

Learning to See the Others

Jane Bennett concludes *Vibrant Matter* with an eloquent manifesto. Indeed, she refers to it as a "Nicene Creed for would-be vital materialists":

I believe in one matter-energy, the maker of things seen and unseen. I believe that this pluriverse is traversed by heterogeneities that are continually doing things. I believe it is wrong to deny vitality to nonhuman bodies, forces, and forms, and that a careful course of anthropomorphization can help reveal that vitality, even though it resists full translation and exceeds my comprehensive grasp. I believe that encoun-ters with lively matter can chasten my fantasies of human mastery, highlight the common materiality of all that is, expose a wider distribution of agency, and reshape the self and its interests. (Bennett 2010a, 122)

This is so eloquent and powerful, yet Bennett is silent regarding indigenous life and presence and intellectual work on this planet. I am struck again and again, reading the new materialisms, by their lack of acknowledging indig-enous people.

The 2001 film *The Others* starring Nicole Kidman comes to mind (dir. Alejandro Amenábar). The protagonist, Grace, lives alone with her two chil-dren in a mansion in the English countryside. World War II rages beyond the quiet stillness of the film. Her husband is away at the front. The chil-dren have a rare disease, xeroderma pigmentosa. Sunlight is poisonous to them. They wake and study and play at night. They sleep in heavily cur-tained rooms during the day. Grace begins to hear noises in the house. The children see spirits. But Grace cannot see those spirits. One night an old woman shows up at the front door with an old man and another young woman. All three were previously servants in the house and offer to come back and help Grace with the children. It eventually becomes clear that Grace, her children, and the servants are all in fact the spirits. Grace will not remember that in a fit of lonely madness she smothered her children and then shot herself. It is the living that Grace cannot see—those invisible presences that make things move in her house. She denies they are there. Yet they terrify her with the claims their presence and movements make on her house. It is the old woman servant who reveals to Grace that they themselves are the spirits, that "the others" are in fact the living. The dead will have to learn to remember their unspeakable acts, make accommodations, live with the living.

Indigenous people, our movements and our voices are the others it seems the new materialists—indeed most of Western thought—cannot fully comprehend as living. They may hear us like ghosts go bump in the night. Once forced to see us, they may be terrified of the claims we make on their house. The invisibility of our ontologies, the very few references to them in their writing, and reference to indigenous thought by other theoretical traditions as "beliefs" or artifacts of a waning time to be studied but not interacted with as truths about a living world—all of this is to deny our vibrancy. It is a denial of ongoing intimate relations between indigenous peoples as well as between us and nonhumans in these lands. We are the living that the new materialists, like so many Western thinkers before them and beside them, refuse to see.

If this theoretical turn is to seriously attend to addressing some of the world's most pressing problems, it needs to learn to see indigenous peoples in our full vitality, not as the de-animated vanished or less evolved. Seeing us as fully alive is key to seeing the aliveness of the decimated lands, waters, and other nonhuman communities on these continents. Understanding genocide in its full meaning in the Americas, for example, requires an understanding of the entangled genocide of humans and nonhumans here. Indigenous peoples cohere as peoples in relation to very specific places and nonhuman communities. Their/our decimation goes hand in hand. Defining and understanding the problem adequately is precisely a situation that requires bridging and reconstituting the relationships between sociality and materiality.

The new materialists may take the intellectual intervention that grounds the vital-materialist creed as something new in the world. But the fundamental insights are not new for everyone. They are ideas that, not so roughly translated, undergird what we can call an indigenous metaphysic: that matter is lively. We Dakota might say "alive." There is "common materiality of all that is": we Dakota might say, "We are all related." That agency should be understood as distributed more widely among human and non-human beings. Understanding these things can help transform how we humans see our place in the world, and therefore how we act. Seeing and understanding matter differently will help us to interrogate the (non)sensibilities of the animacy hierarchy Mel Chen describes, including how such hierarchy shapes the actions of scientists and institutions involved in cryopolitical and other conservationist projects.

When scientists and states conceive of indigenous peoples and lively others as de-animated—the vanishing indigene trope is a constant reaffirmation of this—they assign narrow value to indigenous bodies, histories, and identities. This enables common scientific arguments for the preservation of indigenous biologicals as static storehouses of data or natural resources for the production of knowledge for nonindigenous society. De-animating indigenous peoples enables their domestication and control. And, unfortunately for everyone, it forecloses valuing indigenous peoples and their dynamic conceptual frameworks that could, if taken seriously, help shape broader knowledges and practices in ways that might lessen global devastation.

Notes

1. Oko Iywapi, http://www.wingsprings.com/Oko_Iyawapi/Oko_Iyawapi.html (accessed July 31, 2016).

2. The story is told, among other places, in *Pipestone: An Unbroken Legacy*, a film directed by Sonny Hutchison and Chris Wheeler (2009), available on DVD.

References

Arbour, Laura, and Doris Cook. 2006. DNA on loan: Issues to consider when carrying out genetic research with Aboriginal families and communities. *Community Genetics* 9:153–160.

Bennett, Jane. 2010a. *Vibrant Matter: A Political Ecology of Things*. Durham, NC: Duke University Press.

Bennett, Jane. 2010b. A vitalist stopover on the way to a new materialism. In *New Materialisms: Ontology, Agency, and Politics*, ed. Diana Coole and Samantha Frost, 47–69. Durham, NC: Duke University Press.

Berkhofer, Robert Jr. 1986. *The White Man's Indian, 1820–1880: The Early Years of American Ethnology*. Norman: The University of Oklahoma Press.

Chen, Mel. 2012. *Animacies: Biopolitics, Racial Mattering, and Queer Affect*. Durham, NC: Duke University Press.

Cook-Lynn, Elizabeth. 2012. Indigeneity as a category of analysis. In *A Separate Country: Postcoloniality and American Indian Nations*. Lubbock: Texas Tech University Press.

Coole, Diana, and Samantha Frost, eds. 2010. *New Materialisms: Ontology, Agency, and Politics*. Durham, NC: Duke University Press.

Deloria, Vine Jr. 2001. American Indian metaphysics. In *Power and Place: Indian Education in America*, ed. Vine Deloria Jr. and Daniel R. Wildcat, 1–6. Golden, CO: Fulcrum.

Dippie, Brian W. 1991. *The Vanishing American: White Attitudes and U.S. Indian Policy*. Lawrence: University of Kansas Press.

Eastman, Charles A. 1902. *Indian Boyhood*. New York: McClure, Phillips & Co.

Eastman, Charles A. (1911) 1980. *The Soul of the Indian*. Lincoln, NE: Bison Books.

Eastman, Charles A. 1915. *The Indian Today: The Past and Future of the Red American*. Garden City, NY: Doubleday-Page.

Foucault, Michel. 1990. *History of Sexuality*, vol. 1. New York: Vintage Books.

Foucault, Michel. 2003. *Society Must Be Defended: Lectures at the College de France, 1975–1976*. Trans. D. Macy. New York: Picador.

Genographic Project. n.d. https://genographic.nationalgeographic.com/genographic/about.html (accessed November 30. 2013).

Haraway, Donna Jeanne. 1997. *Modest_Witness@Second_Millennium.FemaleMan _Meets_OncoMouse: Feminism and Technoscience*. New York: Routledge.

Harding, Sandra. 2008. *Sciences from Below: Feminisms, Postcolonialities, and Modernities*. Durham, NC: Duke University Press.

Harris, Cheryl I. 1993. Whiteness as property. *Harvard Law Review* 8:1707–1791.

Human Genome Diversity Workshop 1. 1992. Stanford, CA: Stanford University Press.

Human Genome Diversity Workshop 2. 1992. State College, PA: Penn State University.

Jasanoff, Sheila. 2004. *States of Knowledge: The Co-production of Science and Social Order*. London: Routledge.

Kino-nda-iimi Collective. 2014. *The Winter We Danced: Voices from the Past, the Future, and the Idle No More Movement*. Winnipeg: Arp Books.

Kirksey, S. Eben, and Stefan Helmreich. 2010. The emergence of multispecies ethnography. *Cultural Anthropology* 25 (4): 545–576.

Kowal, Emma. 2013. Orphan DNA: Indigenous samples, ethical biovalue, and postcolonial science. *Social Studies of Science* 43:577–597.

Latour, Bruno. 1999. *Pandora's Hope: Essays on the Reality of Science Studies*. Cambridge, MA: Harvard University Press.

Nadasdy, Paul. 2003. *Hunters and Bureaucrats: Power, Knowledge, and Aboriginal-State Relations in the Southwest Yukon*. Toronto: UBC Press.

Niezen, Ronald. 2003. *The Origins of Indigenism: Human Rights and the Politics of Identity*. Berkeley: University of California Press.

Radin, Joanna. 2012. Taking stock: technologies and ideologies of human tissue preservation in the International Biological Program. Paper presented at Defrost: The Social After-Lives of Biological Substance panel, American Anthropological Association 111th Annual Meeting, November 16, San Francisco, CA.

Radin, Joanna. 2013. Latent life: Concepts and practices of human tissue preservation in the International Biological Program. *Social Studies of Science* 43 (4): 483–508.

Radin, Joanna. 2017. *Life on Ice: Cold War, Frozen Blood*. Chicago: University of Chicago Press.

Reardon, Jenny. 2001. The Human Genome Diversity Project: A case study in co-production. *Social Studies of Science* 31:357–388.

Reardon, Jenny, and Kim TallBear. 2012. Your DNA is our history. *Current Anthropology* 53 (S5): 233–245.

Sagan, Dorion. 2011. The human is more than human: Interspecies communities and the new "facts of life." Paper presented at the Society for Cultural Anthropology's Culture@Large session. American Anthropological Association 110th Annual Meeting, Montreal, October 18.

Shorter, David. 2016. Spirituality. In *The Oxford Handbook of American Indian History*, ed. Frederick E. Hoxie, 1–24. Oxford Handbooks Online. doi:10.109/oxfordhb/9780199858897.013.20.

TallBear, Kim. 2013a. Genomic articulations of indigeneity. *Social Studies of Science* 43 (4): 509–533.

TallBear, Kim. 2013b. *Native American DNA: Tribal Belonging and the False Promise of Genetic Science*. Minneapolis: University of Minnesota Press.

Tigress Productions 2003. *The Journey of Man: The Story of the Human Species*. PBS Home Video.

Waldby, Catherine. 2000. *The Visible Human Project: Informatic Bodies and Posthuman Medicine*. New York: Routledge.

Wolfe, Patrick. 2006. Settler colonialism and the elimination of the Native. *Journal of Genocide Research* 8 (4): 387–409.

Freezing Bodies

10 Suspense: Reflections on the Cryopolitics of the Body

Klaus Hoeyer

On the morning of June 9, 2014, readers of the *New York Times* were confronted with a somewhat paradoxical headline "Killing a Patient to Save His Life." The article described a new experimental treatment for trauma patients who are otherwise likely to die before their wounds can be successfully operated on. "Surgeons will drain their blood and replace it with freezing salt water," readers were told. "Without heart beat and brain activity, the patients will be clinically dead. … By inducing hypothermia and slowing metabolisms in dying patients, doctors hope to buy valuable time in which to mend the victim's wounds" (Murphy 2014). Ethicist Arthur Caplan highlighted the drama of the procedure in his summation: "If this works, what they've done is suspended people when they are dead and then brought them back to life" (Murphy 2014). A life lost, first through trauma and then through clinical suspension, is to be restored. This drama is a drama of suspension in more than one sense: it generates suspense, understood as feelings of worry and excitement, through suspension, that is, delay. The drama furthermore seems to "suspend" more than the time needed to repair patient wounds: the suspension of life and death through hypothermia alters the meaning of life and death. New technologies of this kind also have a tendency to suspend standard bioethical rules. In this case, patients recruited for the experimental procedure will not, for example, be in a position to provide an informed consent. Surgeons plan to recruit victims of gunshot wounds to be trial subjects; according to existing statistics, most experimental subjects are likely to be African-American men. There is an uncanny reminiscence here of the famous Tuskegee study also enrolling African-American men without their consent to generate evidence of use to others. However, the ethical challenges relate to more than reiterations of known social structures and inequalities. Social structures and forms of inequality are not just rehearsed but shaped through these technologies: suspension contains a transformative potential.

Commenting on the ubiquity of media stories of biomedical experimentation, Susan Squier notes that "over our morning coffee we can discover how the foundational categories of human life have become subject to sweeping renegotiation under the impact of contemporary biomedicine and biotechnology" (Squier 2004, 2). One thing Squier does not address, but which is nevertheless central to the changes she explores, is the intersection of *time* and *temperature*. The raison d'être of this volume is to explore the politics at this intersection. It thereby invites us to explore the transformative capacity of not only hypothermia but a range of other technologies operating in the space of, or in the time bought through, cryopreservation. In the following, I will focus on technologies working on human bodies and suggest that suspension through cryopreservation is associated with a heightened sense of drama. These technologies thrive on, generate, and are threatened by a profound sense of suspense.

Control of temperature delivers the ability to "hold still." Radin and Kowal have discussed how "holding still" can be both a supportive act and an act of suppression that detains or constrains life: "The hold of cold can preserve and extend life, or prevent life from being lived" (personal communication with Radin and Kowal, 2013). Cold objects have become understood as "laden with potential"—they are potentialized—and this process opens new avenues for commercialization (Taussig, Hoeyer, and Helmreich 2013). Freezing suspends time and thereby facilitates forms of exchange through which biological material can travel far from its sources and "transgress the desires of [its] descendants" (Radin and Kowal, this vol.). It may sound like a cold-blooded narrative of capitalist expansion, but these moves are entangled in heated dramas relating to technologies aimed at care and enfolded in narratives of what it means to be human, how human bodies can be treated, and fiercely intimate perceptions of self and other. There is a point in using thermal metaphors for emotional engagement, as I just did, because it attunes our attention to the extent to which the technically induced shift in temperature involves socially ingrained motivations and responses. Indeed, cryopreservation delivers a path to a particular form of emotional drama. In the following essay, I suggest that the sense of suspense generated through cooling relates to the ambiguity associated with potentiality. Ambiguity is *productive* in the sense that it allows shifts in meaning (Hetherington and Lee 2000; Svendsen 2011) and generates space for agency (Stark 2009). The cryopolitics of the body unfolds in a space of destabilized definitions and reconfigured relations.

I have already pointed out how the cryopolitics of the body in the case of hypothermia destabilizes the dichotomy of life and death (Bunning, this vol.). In the following I will flesh out another set of dichotomous "foundational categories" that melt through freezing: (1) person and thing and (2) self and other. To do so I review elements of my previous research through the lens of cryopolitics to expose the potentializing effects of temperature control. I argue that potential involves a future-oriented form of ambiguity that invokes semantic shifts and allows new forms of agency to emerge. The suspense generated through suspension reflects a sense of potential that both stimulates and threatens the social entrenchment of medical technologies dependent on cryopreservation.

Person and Thing: The Material Practices of Making Body Parts Exchangeable

Cryotechnologies compose the foundation on which a wide range of biomedical technologies have developed using bits and pieces from human bodies to repair or enhance other bodies: organ transplantation, tissue transplants, regenerative medicine, reproductive technologies, and so on. As noted above and in the introduction to this volume, cooling buys "time," and the time bought facilitates disentanglement of human biological material from the person typically known as "donor." Once disentangled, the material can travel. The cryopreserved disentanglement *potentializes* the material with new possible connections and meanings (Svendsen and Koch 2013).

Traveling necessarily involves socioeconomic infrastructures (Bowker and Star 1999; Star and Griesemer 1989). The fact that material originates in the bodies of *persons* (as far as anything "originates" anywhere and is not just an element of incessant material flows) and is subsequently exchanged in systems dependent on economic compensation has led to intense criticism of what is typically called "commodification" of the human body (Dickenson 2007; Holland 2001; Scheper-Hughes and Wacquant 2002; Sharp 2000). Indeed, cryopreservation can contribute to unsettling forms of exploitation of those without power to resist those who wish to use their body parts. Such technologies thereby extend the grip of socioeconomic inequality to the bodies of the poor who may be forced to sell body parts, or whose body parts are simply taken and sold to wealthier people. Though these modes of exploitation are chilling indeed, I have previously criticized the more general analytical value of commodification theory (Hoeyer 2007, 2013). I will not rehearse all the arguments here, but it is important to note that freezing not only potentializes biological material by making it bio-available (Cohen 2005). The material is also potentialized as "latent life" (Radin 2013), imbued with ambiguous meanings that include a potential for personhood (Hoeyer 2004, 2005). What do I mean by that? A few drops of blood or minor tissue pieces outside bodies would rarely be thought of as representing bits of persons, were they not conserved (held still). Freezing thus preserves the ambiguity of meaning associated with body parts. The potential for association with personhood in turn generates intense regulatory responses that presume the material is different from other objects of trade, or somehow "above" trade altogether (Hoeyer et al. 2009). Accordingly, it is common to talk of, for example, "compensation" or "recovery cost" rather than "price." When once I tried to tease out the micropractices through which a piece of bone nevertheless acquires a price tag, it became obvious that the price-setting process had very little to do with "market forces": there seemed to be an implicit agreement between all involved to avoid such connotations, and the result was far from what would have satisfied *Homo economicus* (Hoeyer 2009). The potential for personhood has real implications for the terms of exchange. It creates a realm of dedicated "unknowing" with regard to the role of money in the exchange.

Even if cryopreserved body parts are rarely treated as plain commodities (at least in the sense of the word suggested by Marx), they cannot remain parts of persons in perpetuity. They transgress distinctions made between persons and things by floating somewhere in between (Fox 2000). Legal systems nevertheless typically build on a basic distinction between those who may own (legal persons, which include companies) and that which can be owned (things, ideas, and nonhuman animals). Cryotechnologies have done much to erode this distinction. To facilitate property relations to cryopreserved body parts, intellectual property law has developed to represent ownership as pertaining only to the work of transforming the body part into a product (Parry 2004). The material itself is said to be beyond trade. In this way, the ambiguous meanings attached to frozen body parts simultaneously imbue them with value and influence their mode of exchange.

The ambiguity of the cryopreserved body engenders a peculiar sense of suspense. The suspension of biological decay creates a space for action in which new social forms are built, new property arrangements emerge, and new hopes and concerns can flourish. Some of these social forms and arrangements inevitably coalesce into scandals of various kinds, as in the case of children's organs kept at Alder Hey Children's Hospital, revelations of illegal trade (Carney 2011), or controversies over cell lines (Skloot 2010). In the hold of cold, hot dreams and chilly nightmares of the future evolve. The depth of the drama reflects not only the legal transgression between person and thing, but also the imaginary space in which ideas about self and other can be transgressed, and it is to this theme that I now turn.

Self and Other: Enacting Boundaries with Cryopreservation

It is part of the myth of modernity that the world arrived at our doorstep in the form of prepackaged knowable and clearly delineable entities (Latour 1993). It is, of course, just an illusion, but a very powerful one indeed. One of the strongest expressions of this modern myth in the bio-medical sciences might be the notion of "a body" as a generic, knowable, universal entity. Farquhar and Lock have called this idea "the singleton model" and pointed to the tendency to align the biomechanical notion with political ideas about personhood and citizenship: "This body proper, the unit that supports the individual from which societies are apparently assembled, has been treated as a skin-bounded, rights-bearing, communi-cating, experience-collecting, biomechanical entity" (Farquhar and Lock 2007, 2). However, as cryopreservation has facilitated transplant and trans-fusion technologies on an altogether new scale, new narratives of body mixtures have emerged.

While cooling allows a heart to travel unaccompanied by the person it used to support, it does not totally sever the ties between this person and the recipient in whom it will resume beating. Anthropologists have col-lected numerous narratives of people experiencing changes in personality after organ transplants, and film and media overflow with dramas thriving on the sense of uncanny mixtures between self and other (Haddow 2005; Lock 2002; Sharp 2006). Challenges to bodily boundaries and the relationship between body and person are not in any way new (Bynum 1995); rather, cryotechnologies have accentuated them in new ways.

It is not just through bodily transfers between persons that cryo-based technologies interfere with perceptions of self and other. During the past decade epidemiology has undergone what is sometimes referred to by my public health colleagues as a "biological turn." Studies of disease now involve not only data gathered in the clinic or elsewhere, but also tissue samples of various kinds. As a consequence, the collection of biomaterial has intensified—and the financial value of existing collections has magnified (Fortun 2008). The ubiquity of tissue sampling implies more than scaling up; it is not just more of the same (Hoeyer 2012). The intensification of biobanking invokes and shifts the political relationship between state and citizen. Access to people's samples has become an arena for negotiating rights and obligations at the nexus of patient, citizen, health-care services, researchers, and state authorities. In the course of such debates the frozen samples themselves have acquired a new potential for agency. To put it bluntly, biobanking has generated a new form of actor in the medical assemblage: the rights-bearing tissue sample. By talking about samples as proxies for persons, freezers become potentialized with intense expectations. Samples stand in for persons in the biomedical sense of the "singleton model," but the political and ethical features of the singleton model then make this imaginary overflow with notions of "persons" being kept on store. Through procedures to renew consent to use old samples, and continuous contact with researchers working on "their tissue," people are invited to think of themselves as distributed in a material sense that works counter to the myth of the singleton model (ironically, of course, per implication of the thinking embedded in the singleton model: that each person has one body, and each body is a clearly identifiable entity with a known identity). In some cases, biobanking furthermore interacts with population-making, as access to stored collections has to be negotiated with the "groups" they are supposed to represent (Kowal 2013; Reardon 2005).

Biobanking, transplants, and regenerative medicine all suspend biological decay and reach out to the future by holding still the biological present. The potentialization of the sample through cryopreservation infuses it with value and with a sense of political danger. Freezing holds still not only the biological process; it also generates the sense of suspense on which dramas of life and death, new perceptions of self, changing understandings of kinship, and fears of commodification thrive.

Concluding Remarks

Cryopreservation is central to most of the technologies of the body currently generating a sense of moral, social, and political suspense. Suspense is what drives a drama, and dramas work on the social settings in which they unfold. The sense of drama makes people—along with things and mixtures of people and things—organize to avoid those mixtures, which for various reasons are deemed dangerous. In the process of defending them, the very meaning of those "foundational categories"—life, death, person, thing, self and other—may shift. Is the hypothermic patient "dead"? Is a piece of bone a "commodity" or a different kind of thing, or no "thing" at all? What is part of a person's "body" and for how long? Is a donated heart part of the recipient or part of the donor? Can a person be distributed as samples in biobanks and as organs living on in other people? What is "self" and "other"?

Cryotechnologies can freeze existing norms, as when African-American men are seen as appropriate trial subjects for hypothermia or when women seek to retain a role as future mothers through egg freezing practices (Inhorn, this vol.); but by "holding still," cryotechnologies also potentialize frozen material with the ability to transform norms. Frozen body parts become disentangled, distributed, reentangled, acquire new meanings, and exert new forms of agency. The exchange of frozen sperm samples can interact with conceptions of personhood and social norms of kinship. Exchangeable organs may interact with familial obligations and the depth of destitution that can be reached through poverty. Tissue grafting and cadaveric organ procurement may interact with changing senses of obligation between the living and the dead. In short, if biology is held still, it is as if other forms of change can be accelerated. Freezing suspends time long enough to facilitate shifts in meaning and changes in social relations, and we are held in suspense while we wait for the drama to unfold.

References

Bowker, Geoffrey C., and Susan L. Star. 1999. *Sorting Things Out: Classification and Its Consequences*. Cambridge, MA: MIT Press.

Bynum, Caroline Walker. 1995. *The Resurrection of the Body in Western Christianity, 200–1336*. New York: Columbia University Press.

Carney, Scott. 2011. *The Red Market: On the Trail of the World's Organ Brokers, Bone Thieves, Blood Farmers and Child Traffickers*. New York: HarperCollins.

Cohen, Lawrence. 2005. Operability, bioavailability, and exception. In *Global Assemblages: Technology, Politics, and Ethics as Anthropological Problems*, ed. Aihwa Ong and Stephen J. Collier, 79. Oxford: Blackwell.

Dickenson, Donna. 2007. *Property in the Body: Feminist Perspectives*. New York: Cambridge University Press.

Farquhar, Judith, and Margaret Lock. 2007. Introduction. In *Beyond the Body Proper: Reading the Anthropology of Material Life*, ed. Margaret Lock and Judith Farquhar, 1–16. Durham, NC: Duke University Press.

Fortun, Mike. 2008. *Promising Genomics: Iceland and deCODE Genetics in a World of Speculation*. Berkeley: University of California Press.

Fox, Marie. 2000. Pre-persons, commodities, or cyborgs: The legal construction and representation of the embryo. *Health Care Analysis* 8 (2): 171–188.

Haddow, Gillian. 2005. The phenomenology of death, embodiment, and organ transplantation. *Sociology of Health & Illness* 27 (1): 92–113.

Hetherington, Kevin, and Nick Lee. 2000. Social order and the blank figure. *Environment and Planning D: Society and Space* 18 (2): 169–184.

Hoeyer, Klaus. 2004. Ambiguous gifts: Public anxiety, informed consent, and biobanks. In *Genetic Databases: Socio-Ethical Issues in the Collection and Use of DNA*, ed. Richard Tutton and Oonagh Corrigan, 97–116. London: Routledge.

Hoeyer, Klaus. 2005. The role of ethics in commercial genetic research: Notes on the notion of commodification. *Medical Anthropology* 24 (1): 45–70.

Hoeyer, Klaus. 2007. Person, patent, and property: A critique of the commodification hypothesis. *Biosocieties* 2 (3): 327–348.

Hoeyer, Klaus. 2009. Tradable body parts? How bone and recycled prosthetic devices acquire a price without forming a "market." *Biosocieties* 4 (2–3): 239–256.

Hoeyer, Klaus. 2012. Size matters: The ethical, legal, and social issues surrounding large-scale genetic biobank initiatives. *Norsk Epidemiologi* 21 (2): 211–220.

Hoeyer, Klaus. 2013. *Exchanging Human Bodily Material: Rethinking Bodies and Markets*. Dordrecht: Springer.

Hoeyer, Klaus, Sniff Nexoe, Mette Hartlev, and Lene Koch. 2009. Embryonic entitlements: Stem cell patenting and the co-production of commodities and personhood. *Body and Society* 15 (1): 1–24.

Holland, Suzanne. 2001. Contested commodities at both ends of life: Buying and selling gametes, embryos, and body tissues. *Kennedy Institute of Ethics Journal* 11 (3): 263–284.

Kowal, Emma. 2013. Orphan DNA: Indigenous samples, ethical biovalue, and postcolonial science. *Social Studies of Science* 43 (4): 577–597.

Latour, Bruno. 1993. *We Have Never Been Modern*. Cambridge, MA: Harvard University Press.

Lock, Margaret M. 2002. *Twice Dead: Organ Transplants and the Reinvention of Death*. Berkeley: University of California Press.

Marx, Karl. 1972. Capital. In *The Marx-Engels Reader*, ed. Robert C. Tucker, 294–437. New York: W. W. Norton.

Murphy, Kate. 2014. Killing a patient to save his life. *New York Times*, June 9. http://www.nytimes.com/2014/06/10/health/a-chilling-medical-trial.html (accessed September 28, 2014).

Parry, Bronwyn. 2004. Bodily transactions: Regulating a new space of flows in "bioinformation." In *Property in Question: Value Transformation in the Global Economy*, ed. Caroline Humphrey and Katherine Verdery, 29–68. Oxford: Berg.

Radin, Joanna. 2013. Latent life: Concepts and practices of human tissue preservation in the International Biological Program. *Social Studies of Science* 43 (4): 483–508.

Reardon, Jenny. 2005. *Race to the Finish: Identity and Governance in an Age of Genomics*. Princeton, NJ: Princeton University Press.

Scheper-Hughes, Nancy, and Loic Wacquant. 2002. *Commodifying Bodies*. London: Sage.

Sharp, Lesley A. 2000. The commodification of the body and its parts. *Annual Review of Anthropology* 29:287–328.

Sharp, Lesley A. 2006. *Strange Harvest: Organ Transplants, Denatured Bodies, and the Transformed Self*. Berkeley: University of California Press.

Skloot, Rebecca. 2010. *The Immortal Life of Henrietta Lacks*. New York: Crown Publishing Group.

Squier, Susan Merrill. 2004. *Liminal Lives*. Durham, NC: Duke University Press.

Star, Susan Leigh, and James R. Griesemer. 1989. institutional ecology, "translations," and boundary objects: Amateurs and professionals in Berkeley's Museum of Vertebrate Zoology, 1907–39. *Social Studies of Science* 19 (3): 387–420.

Stark, David. 2009. *The Sense of Dissonance: Accounts of Worth in Economic Life.* Princeton, NJ: Princeton University Press.

Svendsen, Mette N. 2011. Articulating potentially: Notes on the delineation of the blank figure in human embryonic stem cell research. *Cultural Anthropology* 26 (3): 414–437.

Svendsen, Mette N., and Lene Koch. 2013. Potentializing the research piglet in experimental neonatal research. *Current Anthropology* 54 (S7): 118–128.

Taussig, Karen Sue, Klaus Hoeyer, and Stefan Helmreich. 2013. The anthropology of potentiality in biomedicine: An introduction to supplement 7. *Current Anthropology* 54 (S7): 3–14.

11   The Freezer Program: Value after Life

Jonny Bunning

In 1964, Robert Ettinger was forty-six years old, taught physics at a community college in northern Detroit, and published what he claimed to be the first scientific solution to life's oldest problem. Ettinger had almost succumbed to that problem twenty years earlier, when he was hit by a shell just before the Battle of the Bulge. He survived thanks in part to a new drug named penicillin, then spent four years convalescing in a hospital bed in Michigan. While passing the time, Ettinger claimed, he read an article by a French biologist called Jean Rostand, who had successfully frozen and reanimated frog semen without structural damage. In the process, Ettinger had the idea that would come to define the remainder of his life: that people could similarly be frozen and reanimated. That same idea—its social history and conceptual significance—is the focus of this chapter.

Ettinger has largely been forgotten, but his thesis is surprisingly well known. He believed, as he wrote in his 1964 book, *The Prospect of Immortality*, that cadavers frozen in the present could be revived by improved medicine in the future. Death, he suggested, was not a single threshold, but a series of deteriorating stages which could be reversed through the use of medical technologies. These had increased in power throughout history, were making great leaps forward, and would in all probability continue to do so. Life, its aftermath, and their interface, in other words, were not natural at all. They were technologically contingent, radically historical, and increasingly open to intervention. While all past ages were effectively trapped by the limits of their present, Ettinger implied that new advances in low temperature biology offered the means to exploit historical differences in technical power. He cited studies to show that the time of life was dependent on temperature—the lower the one, the slower the other—so that all living processes, including metabolism and decay, could be suspended for as long as they could be kept sufficiently cold. His solution to the end of life was thus to stop it, not quite living, but not yet fully dead, and hold it still while history ticked steadily onward. When tomorrow's medicine had succeeded in turning today's terminal ills into treatable conditions, he claimed, what currently appeared as a cadaver would be transformed into just one more patient.

In other words, Ettinger's book was a creative blend of scientific discovery and speculative fantasy, held together by a progressive understanding of history. In the mid-1960s, it fit right in. Within a few years the book sold over 100,000 copies and Ettinger reiterated its claims on television, radio, and in magazines as diverse as *Yale Scientific, Christian Century*, and *Paris Match*. Partly as a result, groups of enthusiasts formed to try to put the idea into practice. Against many odds, and unlike some contemporaries on the fringes of science or society, their legacy is tangible to this day. Ettinger, for example, currently hangs upside down in a tank of liquid nitrogen, not far from where he used to teach. He is joined by about two hundred and fifty others—more including disembodied heads and pets—frozen in insulated vats in Detroit as well as Arizona, California, and Russia. Together they form the vanguard of a global cryopolitical movement that challenges many accepted boundaries to life, death, and community. It is a movement that often raises eyebrows and which also raises interesting questions about temporality, value, and scientific salvation.

To engage these questions, I draw on a methodological ethos common in social and historical studies of science for the past thirty years. My aim is not to draw bright lines between legitimate and illegitimate uses of low temperatures, as if their demarcation could be deduced from clear epistemic principles. Nor am I interested in telling stories of the cumulative rise of reason over ignorance, safely consigning these bodies and their freezers to the asylum of failure. Instead, I follow scholars who have pumped credibility back into discarded ideas, accounting for their loss of favor as a contingent (not inevitable) process, dependent on social forces, persuasion, and the enrollment of others.[1] In short, I suspend judgment about the efficacy of cryonics to take the practice and its history seriously.[2]

This approach yields extra benefits because so many people already know something about freezing the dead, most likely because it has enjoyed wide and persistent cultural representation. How many other cryotechnologies have been the basis, *inter alia*, of a Matt Groening cartoon series, a

Woody Allen film, a Don DeLillo book, a Colorado hipster festival, and so many lurid lawsuits? In the United States, the specter of Ted Williams still haunts the practice (Verducci 2003); in France, it was banned after a high-profile trial featuring a home-made freezer, a castle basement, and some deceased parents (Chrisafis 2006). Despite widespread awareness of the practice—or, more accurately, precisely because of it—freezing the dead remains widely misunderstood. Even the name, "cryonics," is often confused with "cryogenics," or the branch of physics dedicated to the study of ultra-low temperatures.[3]

Practitioners of freezing the dead, who call themselves cryonicists, sometimes lament and sometimes denounce these misunderstandings. My goal here is not to simply set the record straight by "following cryonicists around" (see Latour 1987), because their current practice is only the tip of a deeper story of which they themselves are not always the most reliable narrators. Instead of relying solely on participant accounts, I use a range of sources to give a summative (not exhaustive) overview of the history of cryonics. The practice readily entangles science and fiction, and past and future and so my approach is similarly broad, drawing on comics and speculative literature as much as on newspapers or academic articles, in order to trace how the idea moved from narrative strategy in the nineteenth century, to digitally mediated communities of belief by the twenty-first.

This approach is one way in which this chapter differs from two important academic studies of cryonics. Anthropologist Abou Farman (2013), for example, places the practice within a broader spectrum of "American Immortalism," and uses it to explore tensions within the secular cosmology of personhood in its relation to the body. He argues that a Kantian tradition of rationalism, in which a moral subject transcends material causation, forms the heart of secular legal reason, in contrast to a distinct materialist tradition at the root of contemporary science and biomedicine. In liminal cases in which the distinction between person and mere thing is not clear— such as embryos or brain death—the usual pretense of harmony between legal and scientific views of embodiment breaks down, and arbitrary criteria determine the status of the object in question. Cryonics, Farman argues, both exemplifies and strategically navigates this tension. Practitioners rely on the temporally multiple identities of the human body, waiting for it to be medically declared a corpse so that they can treat it as a patient, abbreviating this patient to a brain so that the law will let it move as an organ, yet still refusing to deny subjectivity to the resulting frozen head because (they believe) of its potential for future consciousness. This head is not a thing, Farman argues, but "speculative matter." For this reason, he adds, "the cryopreserved figure of Patient X33 [a disembodied head] emerges as the secular's ideal-typical escape artist" (Farman 2013, 753).[4]

Tiffany Romain (2010) has also examined cryonics with an anthropological eye, and also dwells on its speculative nature, but reads its logic differently. For her, cryonics expresses and amplifies life in the time of neoliberalism, a time characterized by the imperative to optimize and speculate in oneself. This is nowhere more visible than in biomedicine, where investing in the future and investing in the body often become indistinguishable. As an extreme form of such promissory biomedicine, Romain says, cryonics "reveals how speculative economic reasoning is applied to lives and bodies in the United States ... [it is] very much a capitalist form that is similar to other types of insurance and speculative investment" (Romain 2010, 196, 205). In framing the problem this way, she subsumes cryonics into what has become a distinctive literature on "biovalue," in which a Marxian concept of financialization is used to analyze the motivating futurity, economy, and epistemology of much recent biomedical research and development (e.g., Sunder Rajan 2006; Fortun 2008; Cooper 2008; for a useful overview, see Helmreich 2008; for a promising new direction, see Cooper and Waldby 2014).

Despite connecting the speculative nature of cryonics to different sociological dynamics—to secularism; to neoliberal capitalism—both Farman and Romain locate the most developed forms of these dynamics in the United States, so that cryonics (as their concentrated expression) also becomes an exceptionally American cultural practice. Farman, for example, says that it is "part of a complicated post–World War II American history of individualism," that began "and continues to be confined mainly to the United States (a recent center has opened in Russia although there is very little information about it even among cryonicists)" (Farman 2013, 741). Similarly, for Romain, cryonics is a "particularly American social practice, created and taken up by a particular type of American," who is white, male, "geeky," and libertarian (Romain 2010, 196).

Both Farman and Romain discuss cryonics with sophistication, but their approaches risk turning relatively recent features into something more ideal and typical. This makes it difficult to disentangle the rhetorical deployment of history by cryonicists they study from the history of the practice itself. Rather than study cryonics in the present, therefore, I attempt to analyze it historically. In doing so, my goal is less to simply expand an existing map or timeline—although the lack of literature on the topic leads me to discuss the history of the idea in ways that might otherwise be unnecessary—and more to ask how a historically sensitive picture of cryonics allows us to think afresh about time and value. By uncoupling cryonics from any strong connection to the United States and financialized neoliberalism, and so by seeing it as both broader and less cutting edge than sometimes implied, it becomes possible to consider other values of biological materials than those encapsulated by existing theories of "biovalue."

I make this argument in three connected parts. In the first, "History of Cryonics in Fiction and Fact," I look at the genealogy of the idea of freezing the dead, showing its roots in science and fiction, and its transformation into a social movement. This gives a feel for the practice and its changing fortunes, and helps prepare the way for the second part, "History as Cryonics." Here I retain the focus on history, but instead of treating it as a simple container in which to study cryonics, I show how history forms part of the program itself. I suggest that cryonics is oriented not straightforwardly toward the future, but equally toward the past. It posits the present not as a simple 'point' in time, the now, but rather as an intersection where the linear future of the past and the linear history of the future meet. By extrapolating one to the other, Ettinger and his followers built a justificatory schema that runs back from the future, not forward from the present. The attempt to create and then exploit differences in historical and life times, I argue in the third and final part of the chapter, "Realizing the Value of Cryonics," is best described as a form of "vital arbitrage." To develop this notion, I turn not to the smooth growth of finance, but to Marx's analysis of the spasmodic, constipated rhythm that defines the life cycle of capital, in which the value of an object is only visible through a kind of "retro-speculation" from the point of its realization. For these reasons, I suggest at the end of this chapter, the apparently inert Marxist analysis of commodity exchange offers a new and unexpectedly lively way of thinking about frozen bodies and their predicament.

History of Cryonics in Fiction and Fact

When exactly cryonics emerged is hard to say, and raises some problems of definition. Preserving bodies for the afterlife has long been practiced by human societies, and many have been frozen by accident throughout history (Quigley 1998, 2005; Turnbull, this vol.). Speculation that disenchanted science could suspend life or reverse death was not uncommon in the late eighteenth century, and featured in Mary Shelley's *Frankenstein,* a formative work for what would become known as science fiction.[5] The genre's subsequent history has many examples of characters frozen and reanimated later in time. Novels like William Clark Russell's 1887 *Frozen Pirate*, and Louis-Henri Boussenard's 1890 *Dix Mille Ans Dans Un Bloc De Glace* were both popular in the heyday of the "scientific romance," while the 1920s rise of American pulp fiction and the first specifically science-fiction magazine, Hugo Gernsback's *Amazing Stories*, brought many tales that were low in temperature and high in speculation. What distinguishes cryonics from all these precedents, however, is the claim to material, intentional, and science-factual resurrection. That claim was first broached as part of a real amazing story from the same decade.

In 1921, a Siberian Communist, engineer, diplomat, bomb-maker, and former bank robber declared at a friend's funeral, "I am certain that when that time will come, when the liberation of mankind, using all the might of science and technology, the strength and capacity of which we cannot now imagine, will be able to resurrect great historical figures" (in Gray 2011, 161).[6] Three years later, when his former co-conspirator—Lenin—died, Leonid Krasin got his chance. In a strange trinity of Russian Futurism, Soviet Positivism, and Orthodox belief in literal resurrection, Lenin's Funeral Commission was renamed the Immortalization Commission, and hopes rose that Lenin's body could be preserved long enough to give science the chance to resurrect it.[7] Initially, Krasin employed a heating system to keep the corpse at freezing while lying in state during a cold January, but that changed to a cooling system as a mausoleum was hastily constructed for the funeral. Despite best efforts and additional preservatives, however, mold set in, and not even an improved refrigeration unit imported especially from Germany was able to halt the damage. Still optimistic, Krasin made provisions for subterranean refrigeration gear in his brief for the permanent structure to house the body. Eventually, however, a thorough embalming was necessary. Although the plan was ultimately scotched, Soviet scientists continued to invest considerable energy into resurrection.[8]

Contemporary cryonicists, most active in America, tell a different origin story. Benjamin Franklin is frequently quoted as a progenitor, for reasons discussed below, but Robert Ettinger is taken as the true "founding father." An enthusiast of *Amazing Stories* as a child, Ettinger had first set out his own cryonic narrative in a clumsy sci-fi tale, written in hospital and published in the copycat pulp *Startling Stories* in 1948. He eventually went to college on the GI Bill, where he majored in physics and then earned two Master's degrees, one in physics and the other in math. He got a job at a community college. And above all, Ettinger began advocating for a startling idea: "science-factual" cryopreservation as a means to future resurrection, first in a three-page outline sent unsolicited to members of the "Who's Who" list in 1960, then as a sixty-page self-published typescript in 1962. Remarkably, after many rejections, in 1964 the publishing giant Doubleday agreed to publish his ideas as the book *The Prospect of Immortality*. Immortality, Ettinger desperately wanted "scientists and intelligent laymen, and then the nation, and eventually the world," to know, was within reach (Ettinger 1964c, 1).

The basis of that claim is easy to grasp. Ettinger thought that death was not a threshold but a slow slide through a series of stages which could be reversed given the right technology. What were once terminal conditions, he argued, had become entirely treatable with novel techniques like the dialysis machine, cadaveric organ transplants, penicillin or pacemakers, all of which had extended life far beyond its "natural" limits. Similarly, patients whose hearts or breathing had stopped, who had suffered clinical death, could be routinely revived—Ettinger even reported of one Russian scientist who had come back from this kind of death four times (Ettinger 1964c, 43–53). Just as such clinical death had already become reversible, he reasoned, so too would presently more lasting conditions such as biological death (when the brain ceases to function) or cellular death (when cellular structure breaks down) become open to intervention in the future; they would merely be considered illnesses in need of treatment.

Ettinger was calling for the medicalization of the end of life, turning it from a religious or existential problem into a cluster of conditions that were coming unstuck. And in doing so, he was far from the fringe. In the 1960s,

the boundaries between life and death were undergoing one of their periodic reconfigurations, raising a problem for medical professionals charged
with determining the status of bodies that appeared neither fully dead nor
fully alive.[9] While one solution to this problem was to try to stabilize the
distinction into new categories of death, Ettinger saw a more radical possibility. To him, all kinds of death appeared to be in steady retreat, and for
every step he gave a technology, a story, an authority, or a footnote to suggest that it would soon be vanquished entirely. From the perspective
of that victory, Ettinger reasoned, currently dead bodies were simply
patients awaiting treatment ... if they could only be transported forward in
time.

Fortunately, Ettinger revealed to his readers, a new area of research
was emerging, in which biologists were experimentally stopping and
restarting living substances at will (Radin 2017). Ettinger listed a veritable
who's who of the field of cryobiology to support the claim that all kinds
of biological matter could be easily frozen, stored, and revived without
impairment. Ettinger extrapolated from quotations by many leading figures
of the new science—whose experiments involved only small numbers of
cells—that it would soon be possible to cryopreserve whole organs, animals, or even humans. As historian of science Hannah Landecker has
shown, the development of experimental techniques such as freezing
and culturing tissues during the twentieth century allowed the cell to be
turned into "an alienated, exchangeable technical object of our late modern environment," which enabled organisms to "live differently in time"
(Landecker 2005; 2007, 221). Ettinger drew heavily on the possibilities of
the very same examples but concluded, by contrast, that humanity itself
could live differently in time, and that the human body could be transformed into an object capable of exchanging the present for an unalienated
future.

The combined result of this medical and biological redefinition of life,
along with its limits and temporality, was the sober, scientific prospect
of immortality. Or so Ettinger claimed. The book repeatedly argued for
its basis in fact, but equally often slipped into descriptions of interstellar
travel, supersonic transcontinental subways, and other perhaps not totally
substantiated visions:

After awakening, he [i.e. the implicitly white heterosexual male reader of the
Prospect] may already be again young and virile, having been rejuvenated while

unconscious; or he may be gradually renovated through treatment after awakening. In any case, he will have the physique of a Charles Atlas if he wants it, and his weary and faded wife, if she chooses, may rival Miss Universe. Much more important, they will be gradually improved in mentality and personality. (Ettinger 1964c, 6)[10]

Examples like these did not persuade all reviewers, who suspected that Ettinger's book might be slightly less sober than it claimed. Yet they do help locate it in its time. The content of the *Prospect of Immortality* is often uncannily hard to separate from the ambivalent mix of optimism and anxiety that characterized science and society in the early Cold War: some of its technological examples, and much of its outlook, were identical to Vannevar Bush's radiantly optimistic *Science the Endless Frontier* (Bush 1945). Ettinger's estimates of radiation exposure and the need to create underground cryogenic storage vaults to avoid it were eerily similar to contemporaneous military deliberations about freezing animal semen to replenish livestock after a nuclear attack (Hamblin 2013). The central tension in the book—between unstoppable progress and the sense that life is fragile and fleeting, and must be conserved at all costs—is a distinctive product of the Cold War.

Ettinger's *Prospect of Immortality* was so much of its time, in fact, that the same year that he finished an earlier, shorter typescript version in 1962, another man in Washington, DC, self-published a work with a substantively identical claim about the freezer and the afterlife.[11] Evan Cooper, the author of the (pseudonymously published) *Immortality: Scientifically, Physically, Now*, took a more cautious approach to the topic: whereas Ettinger began his book with the assertion that the freezer program "should soon have sledge-hammer impact on every facet of personal and national life," Cooper ended his with an apology and a questionnaire that the reader could send in to refute its thesis. Ettinger believed that the rich or the famous would fund the takeoff of the freezer program before the government stepped in and brought it under the wing of Social Security. Cooper, by contrast, thought that the United Nations was the appropriate authority and recommended an international "cemetery" in the Arctic.

These differences would turn out to be consequential. Believing that the public might not be converted to freezing the dead by the power of the idea alone, Cooper established the Life Extension Society in 1964, and with it a lasting template for amateur immortality. When readers wrote to Ettinger asking for more information, he could refer them to Cooper. And when the government failed to adopt the freezer program as Ettinger originally hoped, advocates shifted toward trying to "make live" again under their own steam. The result was a kind of DIY cryo, a form of networked necrosociality in which those who "suffered" mortality produced their own knowledge and potential cures for their condition, challenging the limits of "life itself" in the process (see Rabinow 1992; Rose 2007). Far from being one of the "sorriest ideas of godforsaken and alienated modernity," as historian Jill Lepore has argued, the forms of death-defying community that emerged were more plausibly a reaction to such alienation, offering sociability to both the dead and the future.[12]

Doubleday, which in 1964 was one of the largest publishers in the United States, pushed the *Prospect of Immortality* in the media, into supermarkets, and onto mail-order book club lists. It also arranged its translation into European languages—Ettinger's book appeared in French the same year it was released in English, then in German and Italian shortly after (Ettinger 1964a, 1965, 1967)—and its promotion as far away as Australia (Ettinger 1964b). As a result, Ettinger's book gained a wide readership among those who believed that they were at the forefront of a world historical change in human existence. An art student in Brooklyn came up with a new, scientific-sounding name for a splinter group—"cryonics"—and soon "cryonics societies" were formed by readers of Ettinger's book in California, Michigan, Florida, and France.

The Cryonics Society of New York held their first annual convention at the New York Academy of Sciences in the spring of 1968, at which a number of issues were discussed by medical doctors, a mortician, a law professor, and other enthusiasts. The Cryonics Society of France, formed the same year, obtained legal permission to build a "cryotorium" in Corsica, chosen because it was cheap and unlikely to suffer a nuclear attack, though it does not seem to have ever been built (Dune 1970).

These new amateur organizations were essential for translating Ettinger's book from second-hand science fiction into a practical project. Yet there was another, more technical dimension to the emergence of cryonics that was equally important. Cryonics could emerge when it did only thanks to the cheap and readily availability of suitable coolants, namely liquid nitrogen and solid carbon dioxide ("dry ice"). By the 1960s, these were already in regular use by cryobiologists and offered the possibility of temperatures far below what could be achieved with mechanical cooling devices (Radin 2017). These coolants eliminated the need for expensive infrastructure of the kind used in Krasin's earlier project; liquid nitrogen could be delivered quite easily by truck, and dry ice could even be transported in the back seat of a family car. More importantly, both substances had dramatically dropped in price thanks to advances in manufacture and storage related to the Cold War (Almqvist 2003). The result was that, in ways never previously possible, very low temperatures were now within reach of the very low-budget DIY-cryo societies. Using a hastily prepared assemblage of dry ice, mortuary drainage pumps, cryoprotectant chemicals, and a modified coffin delivery box, in fact, members of the Cryonics Society of California froze the cadaver of a retired psychologist, Dr. James Bedford, in January 1967 (Nelson 1968). Ettinger's vision was coming to life.

This event generated considerable media attention and implied to some that Ettinger's vision was viable, but it left the more prosaic problem of long term storage unresolved. A wig salesman from Phoenix, Arizona, spotting a new business opportunity, set up a company called Cryo-Care to weld, sell, and even rent steel vacuum cylinders, one of which would soon house Bedford's body. These tanks were, however, prone to leakage and hence expensive to maintain, and sales never took off as hoped. As a result, later in 1967, the New York group started working with a specialist company that made cryogenic semen storage tanks for the Minnesota Valley Breeders Association, and whose work provided the gold standard and prototype for most subsequent vessels (at least until the late 1970s when the company stopped production due to controversy surrounding cryonics). Enthusiasts would use mortuary pumps to replace corpses' blood with cryoprotectant solution, then cool bodies using dry ice, before submerging them in vats of liquid nitrogen. To help retain heat and protect the skin, the bodies themselves were wrapped in aluminum, which, combined with the graphic design of the early cryonics groups, gave an often surreal, space-aged aesthetic to the proceedings.

What members of the nascent cryonics movement really wanted, though, proved elusive. The failure of philanthropic funding or state support was an unfortunate, but not insurmountable, problem; the authoritative validation of scientists and doctors, and perhaps the endorsement of the rich or famous, were far more difficult to manufacture. Some credentialed scientists, such as Columbia University's Gerald Feinberg, were ready supporters, and Ettinger even managed to assemble an advisory committee of scientific "experts" to help with the program, although the qualification for this category was not explained (Mahan 1968). The film director Stanley Kubrick went on record claiming cryonics was "eminently feasible," quoting Ettinger, professing his long interest, and advising a *Playboy* journalist that "within 10 years ... I believe that freezing of the dead will be a major industry in the United States" (Kubrick 2001, 61). Others were less supportive. The thinner the boundary between cryonics and its neighboring fields, in fact, the more work was done to reinforce it. For instance, the Society for Cryobiology, headed by Basile Luyet and formed the year that *The Prospect of Immortality* was published, feared losing its own tentative claims to credibility, especially as the press widely described Ettinger as a scientist and publicized his idea under the banner of "cryobiology" (Radin 2017).

Perhaps less intuitively, cryonicists themselves were engaged in a similar practice of differentiation from those whom *they* saw as liabilities. Some disreputable charlatans, they claimed, were happy to sell the latest prospect of immortality, without any capacity to actually deliver it, a problem apparently aggravated by Ettinger's willingness to put such operators in touch with their clients. As a result, most groups established a noncommercial ethos to demonstrate their legitimacy (Darwin and Platt 2011). Even apparently committed revolutionaries such as Robert Nelson, a TV repairman who had led Bedford's cryonic suspension, were later disowned by others in the movement for being cavalier, opportunistic, and fraudulent. Yet such distinctions between resurrection men were not widely held, and after a few years of initial optimism, the cryonics movement began to lose momentum. Enthusiasts explained their continued lack of acceptance by reference to bad apples, prejudiced scientists, or failure to attract just one more high-profile supporter. But for whatever reason given, the movement did not improve its fortunes. Some very messy legal and storage problems in the 1970s and again in the late 1980s meant that cryonics was soon more likely to appear as the butt of jokes than as the next cutting edge of biotechnology.

Yet the committed few stayed on and doubled down. They produced their own experimental knowledge of freezing and reviving organisms, leading to animal studies and the improvement of "vitrification" techniques to storing bodies, in which substances are ultra-cooled without freezing, producing a solid state like glass. By the early 1990s cryonics membership was rising again, apparently owing more to speculation that new nano and digital technologies would enable the molecular repair and reanimation of bodies in the future rather than any changes in contemporary practice (Drexler 1986; Kurzweil 1992; Parry 2004). In the same period a California court established a helpful legal precedent in the United States, categorizing cryonics as a form of research for which people could anatomically gift their bodies.[13] Further, the 1988 creation of CryoNet, the first cryonics web forum, facilitated the exchange of information and allowed potential converts to bypass what enthusiasts felt was overly biased or misleading news coverage. It was also at this time that a previous distrust of all commerce began to transform into a more complicated relationship to the market, and a partial mimicking of the more mainstream biotechnology boom. Some groups spun off money-making companies to undertake low temperature research that could then support their continually cash-strapped operations; one in the Bay Area came up with the lucrative idea of freezing pets to help the cover costs of their companion species. This was not entirely novel as Ettinger, in his initial comic book version of cryonics, had mentioned that experimental animals would be frozen with his protagonist "for the future biologist to try their skills on" (Ettinger 1946, 108) but the precedent was tactfully set aside.

A similar spirit of innovation still circulates in the early twenty-first century. One of the remarkable things about the desire to freeze the dead, in fact, has been its own persistent plasticity. Just as space travel turned from quasi-communitarian fantasy to X-prize individualism between the early 1970s and 2000s, freezing human bodies too has been largely reinvented from a Cold War vision of mass state-sponsored suspension into an almost singularly technolibertarian venture with ties to a range of other transhumanist groups. While still firmly outside of the mainstream, and likely to stay there for the foreseeable future, the contemporary climate seems more favorable to Ettinger's idea than it has been for some years. A small group of Oxford professors have recently announced their intention to be frozen; a group of sympathetic scientists have defended the practice in the *MIT Technology Review* (Crippen et al. 2015); and the movement may even have finally found the wealthy patrons it has long wished for, Canadian electronics billionaire Robert Miller and casino mogul Don Loughlin (Alsever 2013).

Cryonics as History

The previous section provided historical account of the emergence and endurance of cryonics, which, although necessarily brief, suggests the practice is not best reduced to an expression of neoliberal speculation, nor to an exceptionally American cosmology. Its history appears both more international and more mundane. I want to use this insight to suggest a way to conceptualize the temporality of cryonic freezing without straightforwardly reducing it to sociological accounts of financialization and "biovalue." To understand the logic of cryonics explored in that argument, it helps to briefly pause to look more explicitly at the understanding of time on which the practice relies. History is not just a container in which cryonics happens to be immersed, but a crucial part of the program itself.

It is no coincidence that time travel narratives emerged as a distinct genre in the early nineteenth century, contemporary with the historical novel and, more generally, the rise of history as a story of development (Lukács 1983). If history merely mirrored life and its cycles of birth, growth, and decay—if it was an eternal return of the same—or if history itself were in a state of secular suspension waiting for the Millennium, then being transported into the human future would make little sense, either as a narrative strategy, comedic device, or practical aim.[14] But Ettinger (like Krasin and Bush) saw history rather differently: as a progressive and linear story of human invention, with no foreseeable end.

It is important that cryonicists appear unified in their belief that science too follows this rising path and the history of science plays a fundamental but overlooked role in legitimating the practice. The text of *The Prospect of Immortality* was introduced, quite literally, by this kind of history. The last sentence of the book extends gratitude to "the hosts of the past, our gallant forbears" (Ettinger 1964c, 190). The preface, written by Gerald Gruman, a medical doctor with a recent PhD in history from Harvard University, invokes Alexander Fleming and eighteenth-century "Humane Societies," groups established to revive the apparently drowned.[15] Gruman called Ettinger, in a claim that would be reiterated verbatim and unprocessed by readers, "the latest spokesman for a worthy American tradition going back as far as Benjamin Franklin," because of the latter's musings about being pickled in wine and living to see the future (Gruman in Ettinger 1964c, xiii). The foreword, by Jean Rostand, started not with cryobiology at all, but with a nineteenth-century scientific romance novel about suspended animation. For a manifesto so concerned with medical technology and its future, all this attention to the past seems hard to account for. But although Ettinger's vision was nominally *pro*spective, it relied on a *retro*spective view of the present as already constituting the history of the future.[16] The dominant tense of cryonics is not the simple future, but future perfect: practitioners believe that they always will have been right.

Jean Rostand's foreword to *The Prospect of Immortality* was a preemptive attempt to lend cryobiological legitimacy to the book, and in order to bolster it, the cover prominently identified Rostand as a professor of the Académie Française. And he was. Yet whether American readers knew it or not (the suspicion is that they did not, given how many times this mistake was made in the press), the Académie has little to do with questions of natural science. Its mandate is to govern the correct use of the French language, and its forty members, fittingly, are called "les immortels," after the motto of the Académie, "To immortality." Rostand became an immortal in 1959 after losing the first vote. He was a well-known but iconoclastic biologist who worked on the margins of the establishment and published a staggering eighty-nine books spanning everything from embryology to morality; politics to zoology, eugenics, and history.[17]

Rostand provided Ettinger with more than cryobiological inspiration and a supportive preface. He too had raised the possibility of scientifically transcending death in the 1940s, and speculated about a range of as-yet unrealized biomedical possibilities that passed directly into Ettinger's work, such as the therapeutic use of induced pluripotent somatic stem cells, printing cloned organisms, and much else (Rostand 1959, 14). The idea that such dramatic advances could be imagined but only much later achieved hints at the primary theme of Rostand's historical writing, which portrays past scientists as outstanding but misunderstood luminaries who struggled to gain recognition or were plain ridiculed by their contemporaries. Rostand's view of science, in other words, was firmly in the romantic mode of the genius operating out of joint with contemporary social judgment. In the final volume of his history of biology, for example, he writes:

We have seen, as always, the tenacity of the innovators faced with the incomprehension, if not outright animosity, of those that represent consecrated knowledge. We have seen truth emerge into the light of day slowly, and with difficulty, in despite of the intellectual or sentimental resistance that were created by the respect of prejudice and the attachment to traditions. (Rostand 1945, 235)

The premise for *The Prospect of Immortality* was that technology and mortality were constantly moving through time, so that their meaning (and thus limits) in the present occupied just one position on an upward escalator: "death" meant very different things depending on where one stood. Ettinger, Franklin's heir perhaps, but certainly no inheritor of his diplomatic skills, originally insinuated that the force of history would inaugurate its next stage, the "Freezer Era," leaving detractors in the dust. Yet as it became clear that this alone would not persuade contemporaries, a variation of the same thesis gained importance in which the belated triumphs of past visionaries were overlaid onto the present. From the vantage point of Ettinger's future, the contingency that had characterized life prior to cryonics was narrated as *already* part of history. Robert Nelson, in his programmatic memoire about the first cryonic freezing, gave an example:

Those doctors who have given us their support will one day go down in history, *and a future generation will look back on twentieth-century man* with the same sense of appalled frustration with which we today view seventeenth-century man. ... Each age will have its undiscovered genius, frustrated by his peers, whose exoneration will come only with the passing of time. (Nelson 1968, 33, emphasis added)

Such claims are common in the cryonics literature, and steadily expanded from an attempt to enlist external, professional support to a rhetorical crutch used to keep insiders from losing faith:

So it's not easy being a pioneer cryonicist. Just keep thinking of yourself as the Wright Brothers, as Louis Pasteur, as Galileo. Sure, lots of people called them crazy; but many more people eventually realized they were correct and admired them for their perseverance. (Bridge 1985, 20)

The inferential point of such statements is clear: as with past scientific beliefs, cryonics may appear bizarre or ridiculous now, but its ideas will eventually be rehabilitated—and naysayers judged—by later epochs.[18]

Following Ettinger, cryonicists continue to believe themselves to be the latest avatar of past "misunderstood luminaries" of science, although with one important difference: that if they can remain frozen until their program is rehabilitated, then they will be physically rehabilitated in the process. This move posits the present not as a simple point in time—the now—but rather as an intersection where the future of the past and the history of the future meet. Proponents' insistence on the scientific legitimacy of cryonics is explained by this view, in which the present always appears in the light of the future, and so in which objects and actions do not appear directly ("dead," "strange") but rather as reflected back through other end of the freezer ("patient," "prescient").[19]

Realizing the Value of Cryonics

Cryonics is therefore split into two parts, corresponding to two points in time. There is freezing a corpse on the one, present hand, and thawing and curing a patient on the other, future hand. As noted in the introduction, the temporal aspect of cryonics has been understood as an expression of the logic of finance saturating both biology and everyday American life (Romain 2010). According to this argument, cryonics is merely an extreme form of speculative investment that exaggerates but does not radically differ from the underlying logic of biomedical science, banking, or insurance. That argument effectively subsumes cryonics into studies of "biovalue" and "biocapital" that have emerged over the past decade, themselves drawing more or less directly on Karl Marx (see Anderson, this vol.). And such a framework can indeed be appropriate for understanding biotechnologies like stem cells and, at least to some degree, contemporary biomedicine more generally. Cryonics fits awkwardly into such a theoretical mold, though, and not only because cryonics companies are either not run for profit, or do not make any. While it may be tempting to conflate any idea of the future with contemporary finance, there is another economic logic that is more essentially speculative. The apparently mundane Marxist model of commodity exchange clarifies the logic of cryonics, both economically and epistemically, and, perhaps, reveals a new perspective on "biovalue" that is best suited for the cryopolitical problems presented by "life on ice" (Radin 2017).

Marx is often taken to propose a labor theory of value, a view taken from Ricardo. That is not quite accurate. His understanding of value is based on "socially necessary labor," which can be split into two parts, one corresponding to Ricardo's theory of value, and one adopting Bailey's proto-marginalism. The first aspect of this bivalent theory is the investment of labor to make a commodity, the second is selling it, for which it must meet a social need. Value is only realized if the commodity is sold, rather than being directly contained by the commodity itself (as Ricardo believed). Importantly, selling need not be instant, and there may be a significant delay between the production of a good and its sale, the realization of its value. An unknown painter may spend a lifetime producing a work of art, but never sell it, meaning that the labor invested is not realized. A hundred years later collectors may discover the work and trade it for huge sums, realizing its value, and posthumously turning its producer into an "important" artist. A painting is a relatively stable entity; but consider something more banal and perishable. If somebody invests a certain amount of time and effort to make yogurt, and puts it on the market, then there is a limited period during which it is saleable. After that date, it will degrade, be thrown out, and its value left unrealized. The length of time that such an item can last waiting to be redeemed is based on its material properties, but is not fixed, and can be extended—in this case through refrigeration (see also Woods, this vol.). Similarly, an item can be bought in one place where it is cheap to produce, and sold in another for a profit, or bought, preserved, and sold later when prices have gone up. Exploiting differentials in this way is the basis of arbitrage.

Appropriately, Marx calls the process of putting a commodity onto market a *salto mortale*, translated as a leap of faith but literally meaning a 'deathly leap,' the somersault of a trapeze artist into a void. If the trapeze is not in the right place at the right time, the artist cannot realize the jump: she falls and dies. The risk in the economic sense stems from that fact that money is a universal commodity, and any one commodity is only ever particular. If I have money in my pocket I can exchange it for anything, if I have yogurt then I am in no such position. Says Marx: "The leap taken by value from the body of the commodity into the body of the gold is the commodity's *salto mortale*, as I have called it elsewhere. If the leap falls short, it is not the commodity which is defrauded but rather its owner" (Marx 1976, 200–201). In the case of cryonics, the commodity *is* the owner.[20]

Marx's theory of value, as Kojin Karatani correctly notes, is characterized by a parallax (Karatani 2005): Value is seen first from the point of view of Ricardo, then from Bailey; first from the perspective of production, then from the sale. These two points are connected, but always distinct. The distance between them is indeterminate, and in the intervening period the value embodied by a product is neither present nor absent, but in a state of spectral indeterminacy that Marx called *latent capital*.[21] Given this indirect realization of value, problems arise if no buyer for a commodity can be found—if it is of no use to anyone—as it must be stored and, depending on its specific materiality, may degrade without being valorized.

As is clear, the similarity between such a process and cryonics is significant. Cryonics has a similarly parallax temporality, consisting of preservation and resuscitation. Seen from present, the techniques of cryonic preservation might be considered as a form of extreme experimentation, while resuscitation, which completes the cryonic project, is purely speculative. Similarly, in the present, the cryonic body is valueless, that is to say lifeless. Seen from the alternate position of the future, however, the body becomes a curable (and hopefully cured) patient, meaning that the time and money invested in preserving the body "pays off," while at the same time, the speculative half of *the value of the cryonic project itself is retroactively realized*. The truth value of cryonics, advocates say, can only be seen from this point. There is the obvious problem of why future generations would want to invest in resuscitating corpses, a problem directly analogous to the seller trying to find a buyer for her commodity. Just as commodities are dead labor, in other words, cryonic bodies represent something of a carefully produced dead weight thrown into the marketplace in the hope of finding a buyer to catch them. Thus cryonics is in a very literal sense a speculative "pro-ject," a throwing forward, in the hope that a way of catching frozen bodies will eventually exist and hence that the leap of faith will be retrospectively realized.[22]

There is an important slippage in the language of cryonics. Ettinger argued for the prospect of *immortality*, yet cryonics itself is merely aimed at the prospect of resurrection. Ettinger speculated, as seen above, that if future science was able to "cure" current ideas of death, then it would be equally able to rejuvenate bodies and prevent them from ever totally expiring, which he appeared to believe would occur thanks to stem cells, hormones and molecular biology (Ettinger 1964c, 53–59). The difference is relevant here because, as may already be clear, resurrection follows a rigorous logic of exchange, but immortality follows that of finance. Both are future-oriented, but in distinct ways. The role of finance in an economy is essentially twofold. First, it means that the physical limits of the monetary base can be transcended, meaning that the reproduction of capital need not be constrained by the quantity of, say, gold in circulation. Second, it means that actually putting a commodity onto market—the *salto mortale*—can be deferred. A company that can secure credit may push back selling a new product, effectively converting the money generated from the future sale into present liquidity: once the sale is realized, then creditors are repaid. But a product will be a flop if it fails to attract buyers, in which case the labor invested in its production goes unrealized. If a company was able to secure limitless credit then this risky business of putting a product on the market, could be deferred *forever*. The sale would never be needed to pay back investors.

Immortality follows the same schema. Immortal stem cells, for example, are able to proliferate at will beyond all bodily limits—as the quantity of HeLa cells currently living affirms—while they are similarly able to live beyond the temporal limitations of lifespan. Similarly, the realization of human immortality would mean that the cryonic leap of faith (read: freezing) could be deferred indefinitely through the use of regenerative technologies, affording ever more time, and giving all an indefinite lease of life (read: immortality). Yet the cryonic project itself, while admittedly sometimes slippery in language on this point, is orientated squarely at preserving the bodies of the present and reviving them in the future, with (they hope) no intermediate growths. The dream of transcending bodily limits is common among cryonicists, but shared among trans-humanists more generally, and, I would argue, remains ancillary to and essentially distinct from cryonics, which is concerned primarily with freezing and resurrection, with immortality viewed as a possibility to which future resurrection might merely enable access. Seen in this light, then, we are able to understand why cryonics, while undoubtedly future-focused and highly speculative, cannot be understood as an outgrowth of the financialization of life itself. It also hints at a new modality of analyzing "biocapital," that although one that is inadequate for understanding most forms of regenerative medicine.

Investment in cryonics is, then, speculative. But it is speculative in the sense of selling one's body and self to the future of science, not in the sense of financial speculation. This claim must be elaborated to avoid an easy misunderstanding: the cryonic premise is that one is *already and always* pushed unwillingly into the abyss of death (at least supposing the absence of euthanasia on the one hand and immortality on the other), and that cryonic preservation merely gives the possibility of indefinitely prolonging the plunge (*cadaver*, from *cadere*, "to fall"), and hence the probability of being caught by the future; the probability of cryonicists awaking in the arms and annals of science. The question for cryonicists is hence not "to leap or not to leap," of giving oneself over *or* somehow hanging onto life forever, but rather, that of selling oneself to science or being stolen by death. And on the latter transaction, according to their disenchanted worldview, there is no return.

Similarly, while Romain is correct to note the enhanced "risk management" strategies adopted by cryonicists, including one member who refuses to fly "because, as unlikely as an accident is, a body destroyed in a plane wreck would be impossible to preserve" (Romain 2010, 206), the general category of promise or speculation, assumed to equate to finance, is not the best means of understanding them. Attempts to extend life by following certain diet or vitamin regimens can be understood as a financial activity of sorts in that they allow people to live on borrowed time in the hope that radically regenerative medicine will arrive to extend them unlimited credit before they die—it is worth noting that Ray Kurzweil is both a consummate consumer of neutriceuticals and co-owner of a brand of them (see Ray and Terry's Longevity Products 2012).

To the extent that such risk strategies are aimed not at prolonging life for its own sake but rather at maximizing the quality of the body to be preserved, as in the case of the above infrequent flyer, they are better understood in terms of a constant preoccupation with the exchange value of an object in a way that eclipses its use. Not wearing a pair of shoes so as to be able to sell them for the best price is one example of privileging exchange over use value. Marx gives the example of the miser, whose "passionate chase after value" results in saving rather than using things, making the here relevant note that that English word encompasses both the German *sparen* ("to save") and *retten* ("to rescue") (Marx 1976, 254). In this light, the tendency that Romain notes for cryonicists to avoid danger with a view on their "next" life finds a closer kin in the miser rather than that of the rational investor. As Marx also notes, the capitalist shares the miser's "boundless drive for enrichment," but does so rationally by "throwing his money again and again into circulation" (Marx 1976, 255). The capitalist and the miser both exhibit an ascetic outlook in the interest of profit, but the capitalist's continual leaps mean he is a rational miser, whereas the miser is a perverted capitalist (Marx 1976, 255). The cryonic privileging of the exchange value of life over its use, the risk-averse urge to attain the best convertibility of this life for the next consequently appears here not as a financial strategy so much as—in the strictest sense—a miserably perverted form of life and its value.

Conclusion

Cryonics is approach to life and its aftermath whose core idea involves decoupling biological from technological time, and the exploitation of the difference in technological powers between the present and the future—a kind of vital arbitrage or temporal tourism. As this chapter has shown, that idea has a history that spans science fiction and a Soviet communism, but its contemporary incarnation arose in the United States in the postwar period. It was there that Robert Ettinger, a physics teacher, wrote a book that rather accidentally gained widespread distribution and press attention in several countries, and was translated into a practical program thanks to social organizations on the one hand, and relatively cheap, easily available technical equipment on the other.

Equally sustaining to the cryonic project, and equally overlooked by previous analysts, was a progressive view of history in which medical science (and its social conditions) steadily increased in power, without limit, and in which the present was viewed in an anachronistic manner as always constituting the history of the past. By developing the parallax temporality that characterizes such a perspective, this chapter has showed how the logic of cryonics fits awkwardly into a financial framework, and thus at least hints at a new mode of understanding the value and future-orientation of some other cryotechnological practices. Of course, others may hope that cryonics will provide more value than just this. As one Alcor member and nanotech researcher perceptively put it, "the correct scientific answer to the question 'does cryonics work?' is neither yes nor no. 'The clinical trials are in progress, come back in a century and we'll give you a reliable answer'" (in Delio 2001). Unfortunately, that could be a lifetime away.

Notes

1. Seen in this light, historians look like the more accomplished resurrectors of things long dead.

2. A position that should of course not be confused with advocacy. In fact, if one wanted to argue against freezing the dead, the easier argument rests less on narrowly scientific or technical questions of efficacy—as much as those continue to pose problems for cryonicists—so much as on more classically "social" grounds.

3. To avoid that confusion, but also anachronism, I will use the actors' categories here. Low temperature terminology was unstable at the time that Ettinger's book emerged, and his work was often discussed in the press as a part of "cryobiology." There is good reason to believe that the emergence of cryonics catalyzed other disciplines to standardize their own use of language, with only moderate success. (For a linguist's study of the new language of cryonics see Gordon 1975; for an Oxford cryogenic physicist's account of an attempt to clarify terms at the time, see Kurti 1970.)

4. It is less clear to me that Patient X33 has escaped the problem, so much as deferred its resolution. Farman perhaps recognizes this by talking of "the figure of …" an appropriately indeterminate construction, especially when describing a disembodied head. The question is whether Patient X33 *is* the head or is merely contained in its remains, and with it the problem of mediating the rationalist and materialist split he sees in "the secular." The more interesting tension lies across the idea of the will, both moral (free will), legal (the last will), and temporal (I will) in which the very idea of Western personhood turns out to rest in the possibility of having a future, rather than being confined to the material actuality of the present. Here the secular bind can never be escaped, so much as always avoided by falling forward, as I would suggest Patient X33 is doing for reasons given elsewhere in this essay.

5. Eighteenth-century surgeon John Hunter, based on his theory of "simple life," speculated that "it might be possible to prolong life to any period by freezing a person in the frigid zone," where "he" could be thawed every century and see what had happened in the meantime (Mitchell 2013, 49). This same idea formed the center of an elaborate 1826 hoax, with which Mary Shelley was indirectly involved, centered on a youth who was alleged to have been frozen in an Alpine avalanche for 166 years, thawed, and successfully revived. The story became a news extravaganza in both France and England before being debunked, fueling considerable interest in suspended animation and its feasibility in the process (see Robinson 1975; Mitchell 2013).

6. Russia's first ice-breaker, built in 1916, was later renamed in Krasin's honor.

7. Krasin's influences are discussed at greater length in studies of the so-called Russian Cosmists, who believed in the feasibility of material resurrection and immortality, colonizing other planets and whose sympathizers also included Mayakovsky and Tolstoy (Lukashevich 1977; Young 2012). The "god building" movement, of which Krasin was a part, is also discussed by Nikolai Krementsov (2011, 2014).

8. Lenin's brain was carefully sliced into 30,000 sections, preserved in paraffin wax, and studied for over a decade (Gray 2011, 169). Also of note is that the preservative developed to maintain Lenin's body, declassified only in the 1990s, utilized glycerol long before it became routine in cryobiology (Quigley 1998, 33–35).

9. These forms of indeterminacy interacted, causing trouble for those wishing to harvest organs from ambiguously living bodies (a more common occurrence as more people drove—and crashed—cars). See, for example, debates over "brain death" that took place later in the decade (Ad hoc committee of the Harvard Medical School 1969) and contemporary cross-cultural debates on the subject (Lock 2001).

10. Ettinger's views of consent were questionable, even in his own time. His first cryonics patient was his own mother; when asked if she wanted to be frozen his response was "I don't know if she was really enthusiastic about it, but she was willing." His second patient was his wife. His response to the same question was "She never talked much about it. It was just taken for granted" (both in Lepore 2012, 185).

11. It is notable that Cooper drew heavily on cybernetic theory, while reserving special inspiration from Russian futurist Mayakovsky's 1929 cryosuspension play, *The Bedbug*—"the outline of the method is there" (Cooper 1962).

12. Cryonics even has a romantic streak, at least for some. One practitioner has spoken of falling in love at a cryonics gathering. The pair eventually copublished the *Manrise Technical Review*, instructional manuals aimed at setting cryonics on technically sound foundations (Chamberlain 2012; Chamberlain and Chamberlain 1971). It later became Alcor, now the world's largest cryonics provider.

13. Previously this had been a gray area. An early solution was to build vaults in underground cemeteries where frozen bodies could be stored deeply enough to be legally considered buried. In France, the movement was effectively snuffed out after protracted legal proceedings, bizarre even by the standards of the movement, which resulted in cryonics being deemed an illegal method of disposal for bodies.

14. Robert Mitchell makes a similar observation in his discussion of suspense in the Romantic period and in specific reference to John Hunter's late-eighteenth-century speculations about suspended animation: "Only when social change is understood as endemic, swift, and never ending does it make sense to dream of periodically being frozen and revived in order to 'learn what had happened during [one's] frozen condition" (Mitchell 2013, 53).

15. Gruman's 1966 history of life extension is still cited in the field.

16. For an allied theory of "planned hindsight," see Radin 2015.

17. Rostand translated books by both T. H. Morgan and H. J. Muller into French, and penned one of the very few biographies of Lazzaro Spallanzani, the early modern naturalist who first developed the idea of cryptobiosis.

18. In his chapter on freezing and religion, Ettinger, who was Jewish, briefly discusses Millenarianism, where he hints at the possibility that "the freezer program is part of God's plan," and that it might someday "be accepted as the embodiment of the Millennium" (Ettinger 1964c).

19. Cryonicists belief that any attempt to discredit their practice rests of a prejudiced and insufficiently historical understanding of truth and its confirmation raises compelling problems about the historicity of truth with no easy refutation in authors such as Kuhn (1962), who implies that paradigms are sequential but without discussion of whether or not they might be placed in a more complicated temporal relations—which is effectively what cryonicists are implying: that they are ahead of their paradigm.

20. While there is much to say about the self-reification and retention of personhood in cryonics, I should stress that I see the process not as a simple commercial transaction, but rather as following the same schema of a speculative leap and parallax temporality.

21. Similar to ideas of "latent life" articulated by Carrel (1910) and Keilin (1959) and developed by Radin (2013, 2017).

22. In *The Prospect of Immortality* Ettinger essentially bases redemption on the altruism of future generations, talking of our "friends in the future," but his later shift to Randian egoism leaves this considerable obstacle to the cryonic project curiously unexplored. Cryonics organizations do not reserve money for resuscitations or thawings. If future generations are not indebted to the cryonic project, morally or financially, then their motivating interest in completing the experiment may be less than altruistic, where it exists at all. It is fitting that commodity exchange (unlike finance) implies no existing social relation, which is precisely why (as David Graeber has argued), such transactions have historically taken place at the interfaces between social groupings, rather than within them. While his account looks at this in terms of spatial limits, with cryonics this border is again turned to time.

References

Ad hoc committee of the Harvard Medical School. 1968. A definition of irreversible coma: Report of the ad hoc committee of the Harvard Medical School to examine the definition of brain death. *Journal of the American Medical Association* 205 (6): 337–340.

Almqvist, Ebbe. 2003. *History of Industrial Gases*. New York: Springer.

Alsever, J. 2013. 5 billionaires who want to live forever. *Fortune*, April 4. http://fortune.com/2013/04/04/5-billionaires-who-want-to-live-forever/.

Big sleep, the. 2014. *Economist*, September 6. http://www.economist.com/news/technology-quarterly/21615033-doctors-have-begun-human-trials-suspended-animation-buy-more-time.

Bridge, Steve. 1985. Cryonics in a coonskin cap: Trailblazing the suspension paperwork. *Cryonics* 6 (59): 15–19.

Bush, Vannevar. 1945. *Science: The Endless Frontier*. Washington, DC: US Government Printing Office.

Carrel, A. 1910. Latent life of arteries. *Journal of Experimental Medicine* 12 (4): 460–486.

Chamberlain, L. 2012. Bon voyage Fred Chamberlain. *Chronosphere*. http://chronopause.com/chronopause.com/index.php/page/14/ (accessed December 18, 2012).

Chamberlain, L., and F. Chamberlain. 1971. *Manrise Technical Review*. La Canada, CA.

Chrisafis, Angelique. 2006. Freezer failure ends couple's hopes of life after death. *Guardian*, March 16. http://www.theguardian.com/science/2006/mar/17/france.internationalnews.

Cooper, Evan. 1962. *Immortality: Physically, Scientifically, Now*. Washington, DC: 20th Century Books Foundation.

Cooper, Melinda. 2008. *Life as Surplus: Biotechnology and Capitalism in the Neoliberal Era*. Seattle: University of Washington Press.

Cooper, Melinda, and Catherine Waldby. 2014. *Clinical Labor: Tissue Donors and Research Subjects in the Global Bioeconomy*. Durham, NC: Duke University Press.

Crippen, D. W., R. J. S. Reis, R. Risco, et al. 2015. The science surrounding cryonics. *MIT Technology Review*. https://www.technologyreview.com/s/542601/the-science-surrounding-cryonics/ (accessed January 23, 2016).

Darwin, Mike, and Charles Platt. 2011. Thus spake Curtis Henderson. *Chronopause.com*, February 7. http://chronopause.com/chronopause.com/index.php/2011/02/07/thus-spake-curtis-henderson/.

Delio, M. 2001. Cryonics over dead geeks' bodies. *Wired*, July 20. http://archive.wired.com/culture/lifestyle/news/2001/07/45188.

Drexler, K. Eric. 1986. *Engines of Creation*. New York: Anchor Press/Doubleday.

Dune, Pedro. 1970. La Inmortalidad Por Un Millon de Pesetas? *ABC*, July 26.

Eime, R. 2010. Krassin to the rescue. *North Pole Travel*, June 21. http://www.travelnorthpole.com/2010/06/krassin-to-rescue.html.

Ettinger, Robert C. W. 1948. The penultimate Trump. *Startling Stories*, March, 104–115. http://www.unz.org/Pub/StartlingStories-1948mar-00104.

Ettinger, Robert C. W. 1964a. *L'Homme est-il immortel?* Paris: Denoël.

Ettinger, Robert C. W. 1964b. Scientist says: You can live for ever. *Sydney Morning Herald*, August 30.

Ettinger, Robert C. W. 1964c. *The Prospect of Immortality*. Garden City, NY: Doubleday.

Ettinger, Robert C. W. 1965. *Aussicht auf Unsterblichkeit?* Freiburg im Breisgau: Hyperion.

Ettinger, Robert C. W. 1967. *Ibernazione nuova era*. Milan: Rizzoli.

Farman, Abou. 2013. Speculative matter: Secular bodies, minds, and persons. *Cultural Anthropology* 28 (4): 737–759.

Fortun, Michael. 2008. *Promising Genomics: Iceland and deCODE Genetics in a World of Speculation*. Berkeley: University of California Press.

Frozen Dead Guy Days. 2014. http://frozendeadguydays.org.

Freidberg, Susanne. 2009. *Fresh: A Perishable History*. Cambridge, MA: Belknap Press of Harvard University Press.

Goldstein, E. 2012. The strange neuroscience of immortality. *Chronicle of Higher Education*, July 16. http://www.chronicle.com/article/The-Strange-Neuroscience-of/132819/.

Gordon, W. T. 1975. The vocabulary of cryonics. *American Speech* 50 (1/2): 132–135. doi:10.2307/3087875.

Gray, J. 2011. *The Immortalization Commission: Science and the Strange Quest to Cheat Death*. New York: Farrar, Straus & Giroux.

Hamblin, Jacob Darwin. 2013. *Arming Mother Nature: The Birth of Catastrophic Environmentalism*. Oxford: Oxford University Press.

Helmreich, Stefan. 2008. Species of biocapital. *Science as Culture* 17 (4): 463–478.

Karatani, Kōjin. 2005. *Transcritique: On Kant And Marx*. Cambridge, MA: MIT Press.

Keilin, D. 1959. The Leeuwenhoek Lecture: The problem of anabiosis or latent life: History and current concept. *Proceedings of the Royal Society of London, Series B: Biological Sciences* 150 (939): 149–191.

Krementsov, N. 2011. *A Martian Stranded on Earth: Alexander Bogdanov, Blood Transfusions, and Proletarian Science*. Chicago: University of Chicago Press.

Krementsov, N. 2014. *Revolutionary Experiments: The Quest for Immortality in Bolshevik Science and Fiction*. New York: Oxford University Press.

Kubrick, Stanley. 2001. *Stanley Kubrick: Interviews*. Ed. Gene D. Phillips. Jackson: University Press of Mississippi.

Kurti, N. 1970. Low temperature terminology. *Cryogenics* 10 (3): 183–185.

Kurzweil, Ray. 1992. *The Age of Intelligent Machines*. Cambridge, MA: MIT Press.

Landecker, Hannah. 2005. Living differently in biological time: Plasticity, temporality, and cellular biotechnologies. *Culture Machine* 7.

Landecker, Hannah. 2007. *Culturing Life: How Cells Became Technologies*. Cambridge, MA: Harvard University Press.

Latour, B. 1987. *Science in Action: How to Follow Scientists and Engineers through Society*. Cambridge, MA: Harvard University Press.

Lepore, J. 2012. *The Mansion of Happiness: A History of Life and Death*. New York: Alfred A. Knopf.

Lock, Margaret. 2001. *Twice Dead: Organ Transplants and the Reinvention of Death*. Berkeley: University of California Press.

Lukács, G. 1983. *The Historical Novel*. Lincoln: University of Nebraska Press.

Lukashevich, S. 1977. *N. F. Fedorov (1828–1903): A Study of Russian Eupsychian and Utopian Thought*. Newark: University of Delaware Press.

Mahan, A. F. 1968. Frozen body interred in nitrogen for possible revival. *Owosso Argus-Press* (Michigan), September 5.

Marx, Karl. 1976. *Capital*, vol. 1: *A Critique of Political Economy*. London: Penguin Classics.

Mitchell, R. 2013. *Experimental Life: Vitalism in Romantic Science and Literature*. Baltimore: The Johns Hopkins University Press.

Nelson, Robert. 1968. *We Froze the First Man: The Startling True Story of the First Great Step toward Human Immortality*. New York: Dell.

Nelson, Robert, with Kenneth Bly and Sally Magaña. 2014. *Freezing People Is (Not) Easy: My Adventures in Cryonics*. Guilford, CT: Globe Pequot Press.

Parry, Bronwyn. 2004. *Trading the Genome: Investigating the Commodification of Bio-Information*. New York: Columbia University Press.

Quigley, C. 1998. *Modern Mummies: The Preservation of the Human Body in the Twentieth Century*. Jefferson, NC: McFarland.

Quigley, C. 2005. *The Corpse: A History*. Jefferson, NC: McFarland.

Rabinow, Paul. 1992. Artificiality and enlightenment: From sociobiology to biosociality. In *Essays in the Anthropology of Reason*. Princeton, NJ: Princeton University Press.

Radin, Joanna. 2013. Latent life: Concepts and practices of human tissue preservation in the International Biological Program. *Social Studies of Science* 43 (4): 483–508.

Radin, Joanna. 2015. Planned hindsight: Vital valuations of frozen tissue at the zoo and the Natural History Museum. *Journal of Cultural Economy*. Published online ahead of print http://dx.doi.org/ 10.1080/17530350.2015.1039458.

Radin, Joanna. 2017. *Life on Ice: A History of New Uses for Cold Blood*. Chicago: University of Chicago Press.

Rajan Sunder, Kaushik. 2006. *Biocapital: The Constitution of Postgenomic Life*. Durham, NC: Duke University Press.

Ray and Terry's Longevity Products. n.d. *Brain Health Kit*. http://www.rayandterry .com/.

Rees, Jonathan. 2013. *Refrigeration Nation: A History of Ice, Appliances, and Enterprise in America*. Baltimore: The Johns Hopkins University Press.

Romain, Tiffany. 2010. Extreme life extension: Investing in cryonics for the long, long term. *Medical Anthropology* 29 (2): 37–41.

Rose, Nikolas. 2007. *The Politics of Life Itself: Biomedicine, Power, and Subjectivity in the Twenty-First Century*. Princeton, NJ: Princeton University Press.

Rostand, Jean. 1945. *Esquisse D'une Histoire de La Biologie*. Paris: Gallimard.

Rostand, Jean. 1959. *Can Man Be Modified?* New York: Basic Books.

Rostand, Jean. 1960. *Discours de Réception à l'Académie Française et Réponse de Jules Romains*. Paris: Gallimard.

Seung, S. 2012. *Connectome: How the Brain's Wiring Makes Us Who We Are*. Boston: Houghton Mifflin Harcourt.

Verducci, T. 2003. What really happened to Ted Williams. *Sports Illustrated*, August 18. http://www.si.com/vault/2003/08/18/348299/what-really-happened-to -ted-williams-a-year-after-the-jarring-news-that-the-splendid-splinter-was-being -frozen-in-a-cryonics-lab-new-details-including-a-decapitation suggest-that-one-of -americas-greatest-heroes-may-never-rest-in.

Waldby, Cathy. 2002. Stem cells, tissue cultures, and the production of biovalue. *Health* 6 (3): 305–323.

Young, G. M. 2012. *The Russian Cosmists: The Esoteric Futurism of Nikolai Fedorov and His Followers*. New York: Oxford University Press.

12   The Frozen Archive, or Defrosting Derrida

Warwick Anderson

In October 2004, the French philosopher Jacques Derrida was buried in a cemetery outside Paris. At the age of seventy-four, he died from pancreatic cancer (Peeters 2012). Derrida wanted his corpse to be buried—a process that seemed to him "less inhumane than cremation" (Derrida 2011, 160). That way he would not instantly be annihilated; his disappearance would take time. "As a buried corpse, I would still have a place reserved to me, I would have a proper place, I could still take place" (Derrida 2011, 161; see also Naas 2012). Cremation, if decided by the dying, seemed to him "a sort of irreversible suicide," an obliteration that made the "labor of mourning both infinite and null" (Derrida 2011, 162). Derrida weighed up burial and cremation as he faced his own death and as he prepared for the funeral of his friend Maurice Blanchot, who had prescribed cremation. It was a deeply felt philosophical problem. For Derrida, all thought of death was inextricable from thought of survival, the survival of the remainder, the remains: thus funeral procedures "deliver the corpse over to its future" (Derrida 2011, 132). But who is the other that takes charge of the remainder? What is it, and who is it, that makes a thing out of the once living? What becomes the value of this thing, the remainder?

Toward the end of his life Derrida was preoccupied with the cultures of the corpse, whether earth bound or fire borne. Yet freezing, contrary to cremation, seems not to have crossed his mind. The thought of a *frozen* Derrida is intriguing. Since World War II, cryobiologists have been studying suspended animation, making "latent life" or "anabiosis," trying to show that "life may be a discontinuous process," hoping to immortalize tissues in breach of death's dominion (Keilin 1959, 181, 186).[1] Sadly, none of Derrida's tissues were frozen, so far as I know—not even his brain. But in this essay, perversely, I still want to defrost, and mobilize, Derrida. I wish to raise the specter of Derrida: to have his semicorporeal form, his latent life, haunt the discussion of value, to show how a haunting gives value (Derrida 1994, 212).[2]

But first, let's step back from the philosopher's burial site a moment to trace the passage of frozen brains from the eastern highlands of New Guinea to the National Institutes of Health in Bethesda, Maryland. In the 1960s, scientist D. Carleton Gajdusek was accumulating the brains of Fore people who died of the neurological disease kuru, a mysterious epidemic that had killed almost a third of this isolated group. Fresh brains were flown across the Pacific to his laboratory to be inoculated in chimpanzees; after a few years the animals succumbed to kuru, proving the disease was transmissible and earning Gajdusek his 1976 Nobel Prize (Anderson 2008, 2013). To preserve the brains on their rapid transit, investigators used dry ice (liquid carbon dioxide) and Styrofoam containers, keeping tissues frozen at around minus 80 degrees Celsius. Dry ice was patented in 1925 and Dow Chemical began making Styrofoam in the 1940s, but both products became popular in the 1950s—during the Cold War, of course. Adept at preserving and mobilizing the tissues of his beloved "primitives," Gajdusek became obsessed with maintaining the trans-Pacific cold chain.[3] Much later, he told me that even if they had stuck the brain tissue in a coat pocket and traveled for a year, it still could have transmitted the slow virus, or prion (infectious protein)—the putative cause of the fatal brain disorder. But through the 1960s, Gajdusek did not come in from the cold.

In early November 2012, after Hurricane Sandy struck, the *New York Times* reported: "The calls started coming in late Tuesday and early Wednesday: offers of dry ice, freezer space, coolers" (Carey 2012). According to the article, medical researchers in New York were seeking to save their biological specimens:

Staff members at NYU worked around the clock to preserve research materials, running in and out of darkened buildings without elevator service, hauling dry ice and other supplies up anywhere from two to more than 15 floors. … Susan Zolla-Pazner, director of AIDS research at the Manhattan Veteran Affairs Medical Center, had lost power in her first floor lab at 23rd Street and First Avenue. She finally hired a company to haul her 20 freezers, full of specimens, for safekeeping.

How vulnerable the brief dry-ice age of biomedicine is to the effects of anthropogenic global warming. How fragile, or mortal, the cold chain becomes in an overheated world. "We spent all of Tuesday and Wednesday hauling 1300 pounds of dry ice up to the 18th floor, using the stairs, to stabilize the freezers first," reported Dr. Zolla-Pazner.

Spectral Commodities

Of course, freezing technologies continue an older scientific project of fixing or stabilizing biological samples and rendering them mobile—or as Bruno Latour puts it, making "immutable mobiles" (Latour 1983). It also is the history of trying to de-animate and forget persons, to turn donors into future things, immortal things, to accumulate novel collectivities with artificial modes of association or biosociality—in other words, to ontologize the remains, to make present the partible. It is worth asking what, or who, constitutes the value of this frozen stuff to the warm living, whether scientists, kin, Indigenous people, venture capitalists, or even critical scholars of science.

Anthropologist Webb Keane writes of the "tension between epistemologies of estrangement and of intimacy" (Keane 2005, 62; see also Megill 1994). Analysts of the cultures of modern science tend to favor intimacy and entanglement, and to suspect or resist objectification. We romantics don't trust the frozen; it leaves us cold. At the same time, the estrangement of frozen things allows scientists to speak for—and with—them. Kin and indigenous people are left to speak for prefrosted subjectivity and defrosted debris. But surely there is a more complicated dialectic between frozen objectivity and warm subjectivity. An ideal, or abstract, scientific trajectory would trace an arc from primitive, or belated, persons toward things serviceable in the future, thereby moving us on from the past and bypassing the contemporary. Accordingly, freezing should result in temporal displacement as well as spatial mobility. But have we ever become thoroughly objective? Isn't it more likely we crash into an archive of biomedical debris, into the postcolonial museum of the ruins of biomedicine?

Let me use kuru again to consider the transformation of persons into things, of the sacred, or "unscathed" as Derrida puts it—that which allows no, or ritually limited, circulation—into the profane, freely mobilized, objects of technoscience (Derrida 1998). I say "transformation" but really I mean sorcery, or conjuring. Anthropologist Shirley Lindenbaum describes Fore sorcery as an illicit attempt to interrupt and rechannel the flow of goods, to circumvent conventional exchange relations and reverse value, thereby upsetting social structure and hierarchy (Lindenbaum 2013, 65). Anything intimately connected with an intended victim might become an ingredient in sorcery, any bodily discards or leavings, especially feces. Such materials, along with a "poisonous" stone, would be wrapped in leaves and placed on swampy ground. In this process, the spell and performance were crucial techniques, as they reinvigorated and by projection reinserted the poisoned bodily discards, to cause damage in the living. As the bundle decayed, the victim weakened until death occurred.[4]

Similarly, scientists can take a personal material such as blood or tissue, freeze it, and remove it to a laboratory, sometimes even to the post-Sandy swampiness of the labs of New York. In performing normal science, they endeavor to make human tissues scientifically serviceable and to make the person, the donor, disappear—to perform a magical exorcism. Fore sorcerers were turning "things"—as they appear to us—into live persons, so their bundles might have effects on contemporary victims. Scientists, in contrast, are seeking to make persons into frozen things—into scientific valuables, suspended forms of life no longer commensurate with estimates of donor personhood. They seek to make "who" into "what." The deep freeze of the laboratory thus is a place of modern conjuring and enchantment, as much as enlightenment. The crux of all this transformational movement, the point of greatest difficulty, the moment when modern magic is required, is when scientists try to cut one network, to use Marilyn Strathern's phrase, and make possible another network, a scientific network (Strathern 1996). This disruption, which is at the same time a moral deviation or reorientation, is at least as important as the obvious spatial and temporal mobilities involved. As Derrida puts it in *Specters of Marx*, such conjuration is "first of all an alliance"; it is "a matter of neutralizing a hegemony or overturning some power" (Derrida 1994, 58).

But listen to Carleton Gajdusek, the scientist, as he holds the brain of a Fore child in 1959: "How strange to fondle the brain of the infant I so well remember, of the mother I shall never be unable to recall" (Gajdusek 2002, 39–40). His specimens, so arduously acquired in fraught exchanges, retained, despite mechanical reproduction, the aura of the person from whom they came (Benjamin 1968). For Gajdusek, the value of these tissues was constituted with reference to a haunting, to what Derrida calls the "spectral effect of the commodity" (Derrida 1994, 193). Spectrality consists of the apparition of the inapparent, the paradoxical incorporation of the body that was supposedly effaced. "The fetish," Derrida writes, "would be the given, or rather lent, borrowed body, the second incarnation conferred on an initial idealization, the incorporation in a body that is, to be sure, neither perceptible nor invisible, but remains flesh, in a body without nature, an *a-physical* body that could be called, if one could rely on these oppositions, a technical body or an institutional body" (158). In spite of laboratory purification or exorcism, this spectral body, again according to Derrida, "accumulates undecidability, in its uncanniness, their contradictory predicates: the inert thing appears suddenly *inspired*, it is all at one transfixed by a *pneuma* or a *psyche*" (192).[5] Or, as Gajdusek mourned, "how strange to fondle the brain of the infant I so well remember." The scientist, in Derrida's words, "hesitates between the singular 'who' of the ghost and the general 'what' of the simulacrum" (212). This sensuous or phantomatic materiality seems to me what Michel Serres means when he refers to the proliferation of "quasi-objects" (Serres 1982), and what Latour means when he asserts: "We have never been modern. Modernity has never begun" (Latour 1993, 47).

Of course, I am talking about the deconstructive logic of science, the internal reactivity of all forms of sovereignty, and the duplicity of sources, in general. In other words, I am talking of the need in science for the "testimonial signature"—and the "animist relation to the tele-technoscientific machine" (Derrida 1998, 45, 56). As Derrida (1998, 45) observes in his essay "Faith and Knowledge":

The scientific act is, through and through, a practical intervention and a technical performativity in the very energy of its essence. And for this very reason it plays with place, putting distances and speeds to work. It delocalizes, removes or brings close, actualizes or virtualizes, accelerates or decelerates. But wherever this tele-technoscientific critique develops, it brings into play and confirms the fiduciary credit of an elementary faith. ... We speak of trust and of credit or of trustworthiness in order to underscore that this elementary act of faith also underlies the essentially economic and capitalistic rationality of the tele-technoscientific. No calculation, no assurance will ever be able to reduce its ultimate necessity, that of the testimonial signature.

For Derrida, then, what he calls tele-technoscience is expropriative and delocalizing, denying traditional markers of identity at the same time as it relies on them, embracing bodily markers of identity (Derrida 1998).

Primitivism and modernity thus facilitate and validate each other. Indeed, technoscience, through its intrusive sampling techniques, enables new primitives and produces a past it can theatricalize as its own difference. Additionally, the collecting of specimens from the living, the rendering of the living as inanimate, surrendering to the Freudian death drive—to technics, the machine, the prosthesis—opens the community to something other, something more than itself (Freud 1922). For Derrida, this auto-immunitary process, the *"autoimmunity of the unscathed,"* which he associates with technoscience, opens us up to the future (Derrida 1998, 80). It serves life by separating it from itself, and making it iterable, unlively. "With absolute immunity," Derrida writes, "nothing would ever happen or arrive; we would no longer wait, await, or expect, no longer expect one another, or expect any event" (Derrida 2003, 152; see also Derrida 2005). Thus the Fore people imagined technoscience—albeit technoscience interpreted as another method for countering sorcery—as opening a path for them to become modern people; Gajdusek needed modern technoscience, the cross-reactivity of the objective and subjective, to discover and sacralize, or indemnify, his primitives. In the mixed economy of Gadjusek's object world, there was no distinction between those whom W. H. Auden called "the tender who value" and "the tough who measure" (Auden 2003, 32).

In 1944, Auden was giving voice to Caliban, who stands in and speaks for the dead author of *The Tempest*, William Shakespeare. For Auden, *The Tempest* was essentially a meditation on death, a reflection on the paradoxes of art and life, with Caliban representing flesh, the animalistic, and Ariel an insubstantial, ethereal, nonhuman presence. Throughout the play, Prospero uses magical spells to counter death and to enslave Caliban. But Auden lets Caliban speak elegantly, revealing the false coinage of the play's dualism. He suggests that life is latent in art, that art remains haunted by flesh (Auden 2003).

The Frozen Archive

What does this mean for the memory work of the specimen archive? Gajdusek's kuru brains and other Fore tissues now exist—if that is the term—as a scattered frozen archive of biomedical debris. According to Michel Foucault in *The Archaeology of Knowledge*, the archive is

that which determines that all these things said do not accumulate endlessly in an amorphous mass, nor are they inscribed in an unbroken linearity, nor do they disappear at the mercy of chance external accidents; but they are grouped together in distinct figures, composed together in accordance with multiple relations, maintained or blurred in accordance with specific regularities; that which determines that they do not withdraw at the same pace in time, but shine, as it were, like stars, some that seem close to us shining brightly from far off, while others that are in fact close to us are already growing pale. (Foucault 2002, 145–146)[6]

In *Les lieux de mémoire*, historian Pierre Nora claims: "The obsession of the archive is a mark of our times" (Nora 1984, xxvi).

Human specimens require an archive in order to maintain functionality over a period greater than the longevity of their referents—or their reverence. The specimen needs tight links to provenance, explicit registration, and a complete index. Access to the institutional archive is limited, circulation of the material is restricted, and personal information is treated as confidential. But what can the archived specimen *do*? Or rather, what does *archiving* do for the living? According to Derrida, "The technical structure of the *archiving* archive also determines the structure of the *archivable* content even in its very coming into existence and its relationship to the future. The archivization produces as much as it records the event" (Derrida 1996, 17). Moreover, he argues, we have no fixed concept of the archive, only an impression: "an insistent impression through the unstable feeling of a shifting figure, or of an in-finite or indefinite process" (19). The disruptive force of deconstruction is contained within the architecture of the archive. Even as Gajdusek's profoundly hypothermic colonial archive appears to ontologize remains, or exorcize and expel subjectivity, it remains haunted, a spectral archive. No degree of scientific rigor will conjure away the spirits of the donors—nor, for that matter, the spirits of its collectors. In the end, despite everything, that is the value of the collection.

In a different context, Ann Laura Stoler points out that archiving as a process is at least as revealing as the archive is as a thing, or collection of things. According to Stoler, colonial archives "were both transparencies on which power relations were inscribed and intricate technologies of rule in themselves" (Stoler 2009, 20). She urges us to treat the archive, regardless of its contents, "as a force field that animates political energies and expertise, that pulls on some 'social facts' and converts them into qualified knowledge, that attends to some ways of knowing while repelling and refusing others" (Stoler 2009, 22; see also Dirks 2002). One might read into Stoler a postcolonial argument for the need to reanimate the archive, to counter the conjuration of ontology, or perhaps simply to appreciate the specters that already inhabit it, to whom it remains hospitable. As Derrida puts it, "The space of such technical experience tends to become more animistic, magical, mystical" (Derrida 1998, 56). All one needs to do to deconstruct is to do memory work.[7]

Defrosting Derrida

What, then, is the value of latent life? Or who is the value of latent life? In the face of death, as the result of necropolitics, these specimens, these remains, might offer us a biopolitical future. Potentially, these things can be reanimated or inspired again, extending life, or at least bringing a spectral presence back into life (Taussig, Hoeyer, and Helmreich 2013). As anticipatory remainders, or reminders, they dwell in the future (Adams, Murphy, and Clarke 2009). Derrida frequently, toward the end of his life—what he thought to be the end of his life—came back to scene of cannibalistic consumption, the devouring of the body, which he associated with being buried alive, waking in a tomb. The buried body eventually rots and decomposes; the earth consumes it. "Might sovereignty be devouring?" he asked. "The place of *devourment* is also the place of what carries the voice" (Derrida 2009, 23). The site of objectification and consumption, it becomes the place where one opens the self, or it is opened, to the future. Until, that is, the archive of latent life, or the cemetery, turns into a ruin—which is to say, when our subjective investments decay.

Burying, burning, freezing, eating—all are different processes for dealing with the dead, for managing human remains or partible tissues. Even as we conventionally use such rituals to make sure persons—distinct from the things they become—do not come back or hang around, some incalculable trace, some spectral presence, may linger. Tissue freezing is supposed to help us forget or destroy the person, to make us safe from revenants, to stabilize objects, but in so doing it also can render unsafe, or keeps lively, the stuff it produces, thus lending value to the remainder. This haunting of the scientists' specimens is the hope of cryonics enthusiasts, even if "the status of the cryonic subject is indefinitely deferred" (Doyle 2003, 68). Indeed, while tissue cryopreservation tries to produce things even though they might retain spectral personhood, cryonics tries to render persons latent yet ends up with high-maintenance things (Bunning, this vol.). As a result, "cryonics is not so much an attempt to avoid the great lack of death as the purchase of a certain style of death, a death of some difference" (Doyle 2003, 77; see also Farman 2013). These processes, involving disappearance and return, and abiding death, in Sigmund Freud's terms, are "making what is in itself disagreeable the object of memory and psychic preoccupation" (Freud 1922, 16). No wonder there is something melancholic about the frozen archive.

Derrida could have had more to say about the melancholic potential and lingering value of latent life. Perhaps it was derelict of him not to consider the thermodynamics of biovalue—not to consider his being frozen.

Notes

1. *Anabiosis* is derived from the Greek term meaning "coming back to life." On the Californian mode of cryogenic storage, see Ettinger 1965. See also Parry 2004 and Radin 2013.

2. As Stefan Helmreich (2011, 695) claims, "A good case could be made that today's biopolitics are ever more entangled with necropolitics" (see also Agamben 1998; Esposito 2008). In part, this essay serves as a response to Radin's suggestion that I should examine the contribution to the creation of value in science of freezing technologies as well as tissue exchange. More generally, it serves as a supplement to more economistic accountings of biovalue: see Waldby and Mitchell 2006, Sunder Rajan 2006, and Cooper 2008.

3. Radin (2013, 488) argues that collectors in the 1960s regarded the frozen tissues of isolated people as a "resource for a future in which primitive peoples no longer existed."

4. I am grateful to Shirley Lindenbaum for clarifying aspects of kuru sorcery to reinforce my argument.

5. Thus "ontology opposes it only in a movement of exorcism. Ontology is a conjuration" (Derrida 1994, 202).

6. See also Arvatu 2011.

7. "Since the disruptive force of deconstruction is always already contained within the architecture of the work, all one would finally have to do to be able to deconstruct, given this *always already*, is to do memory work" (Derrida 1989, 73).

References

Adams, Vincanne, Michelle Murphy, and Adele E. Clarke. 2009. Anticipation: Technoscience, life, affect, temporality. *Subjectivity* 28:246–265.

Agamben, Giorgio. 1998. *Homo Sacer: Sovereign Power and Bare Life*. Trans. D. Heller-Roazen. Stanford: Stanford University Press.

Anderson, Warwick. 2008. *The Collectors of Lost Souls: Turning Kuru Scientists into Whitemen*. Baltimore: The Johns Hopkins University Press.

Anderson, Warwick. 2013. Objectivity and its discontents. *Social Studies of Science* 43:557–576.

Arvatu, Adina. 2011. Spectres of Freud: The figure of the archive in Derrida and Freud. *Mosaic* 44:141–159.

Auden, W. H. 2003. Caliban to the audience. In *The Sea and the Mirror: A Commentary on Shakespeare's The Tempest*, ed. Arthur C. Kirsch. Princeton, NJ: Princeton University Press.

Benjamin, Walter. 1968. The work of art in the age of mechanical reproduction. In *Illuminations*, ed. Hannah Arendt, 214–218. London: Fontana.

Carey, Benedict. 2012. A collective effort to save decades of research as the water rose. *New York Times*, November 6. http://www.nytimes.com/2012/11/06/health/a-collective-effort-to-save-decades-of-research-at-nyu.html.

Cooper, Melinda. 2008. *Life as Surplus: Biotechnology and Capitalism in the Neoliberal Era*. Seattle: University of Washington Press.

Derrida, Jacques. 1989. The art of memories. Trans. Jonathan Culler. In *Memories for Paul de Man*, rev. ed., 45–88. New York: Columbia University Press.

Derrida, Jacques. 1994. *Specters of Marx: The State of the Debt, the Work of Mourning, and the New International*. Trans. P. Kamuf. New York: Routledge.

Derrida, Jacques. 1996. *Archive Fever: A Freudian Impression*. Trans. E. Prenowitz. Chicago: University of Chicago Press.

Derrida, Jacques. 1998. Faith and knowledge: The two sources of "religion" at the limits of reason alone. In *Religion*, ed. Jacques Derrida and Gianni Vattimo, 1–78. Cambridge: Polity Press.

Derrida, Jacques. 2003. Autoimmunity: Real and symbolic suicides. Trans. P.-A. Brault and M. B. Naas. In *Philosophy in a Time of Terror: Dialogues with Jürgen Habermas and Jacques Derrida*, ed. Giovanna Borradorri. Chicago: University of Chicago Press.

Derrida, Jacques. 2005. *Rogues: Two Essays on Reason*. Trans. P.-A. Brault and M. B. Naas. Stanford: Stanford University Press.

Derrida, Jacques. 2009. *The Beast and the Sovereign*, vol. 1. Trans. G. Bennington. Ed. Michel Lisse, Marie-Louise Mallet, and Ginette Michaud. Chicago: University of Chicago Press.

Derrida, Jacques. 2011. *The Beast and the Sovereign*, vol. 2. Trans. G. Bennington. Ed. Michel Lisse, Marie-Louise Mallet, and Ginette Michaud. Chicago: University of Chicago Press.

Dirks, Nicholas B. 2002. Annals of the archive: Ethnographic notes on the sources of history. In *From the Margins: Historical Anthropology and its Futures*, ed. Brain Keith Axel, 47–65. Durham, NC: Duke University Press.

Doyle, Richard. 2003. Disciplined by the future: The promising bodies of cryonics. In *Wetwares: Experiments in Postvital Living*, 62–86. Minneapolis: University of Minnesota Press.

Ettinger, R. C. W. 1965. *The Prospect of Immortality*. New York: Sidgwick & Johnson.

Esposito, Roberto. 2008. *Bios: Biopolitics and Humanity*. Trans. T. Campbell. Minneapolis: University of Minnesota Press.

Farman, Abou. 2013. Speculative matter: Secular bodies, minds, and persons. *Cultural Anthropology* 28:737–759.

Foucault, Michel. 2002. *The Archaeology of Knowledge*. Trans. A. M. Sheridan-Smith. London: Routledge.

Freud, Sigmund. 1922. *Beyond the Pleasure Principle*. Trans. C. J. M. Hubback. London: International Psycho-Analytical Press.

Gajdusek, D. Carleton. 2002. *Journal of Continued Quest for the Etiology of Kuru with Return to New Guinea, January 1, 1959–December 31, 1959*. Gif-sur-Yvette, France: Institut de Neurobiologie Alfred Fessard.

Helmreich, Stefan. 2011. What was life? Answers from three limit biologies. *Critical Inquiry* 37:671–696.

Keane, Webb. 2005. Estrangement, intimacy, and the objects of anthropology. In *The Politics of Method in the Human Sciences: Positivism and Its Epistemological Others*, ed. George Steinmetz, 59–88. Durham, NC: Duke University Press.

Keilin, D. 1959. The problem of anabiosis or latent life: History and current concepts. *Proceedings of the Royal Society of London, Series B: Biological Sciences* 150:149–191.

Latour, Bruno. 1983. Give me a laboratory and I will move the world. In *Science Observed*, ed. Karin Knorr and Michael Mulkay, 141–170. London: Sage.

Latour, Bruno. 1993. *We Have Never Been Modern*. Trans. C. Porter. Cambridge, MA: Harvard University Press.

Lindenbaum, Shirley. 2013. *Kuru Sorcery: Disease and Danger in the New Guinea Highlands*. 2nd ed. Boulder: Paradigm.

Megill, Allan, ed. 1994. *Rethinking Objectivity*. Durham, NC: Duke University Press.

Naas, Michael. 2012. To die a living death: Phantasms of burial and cremation in Derrida's final seminar. *Societies* 2:317–331.

Nora, Pierre. 1984. Entre mémoire et histoire: la problématique des lieux. In *Les lieux de mémoire. Tome 1: La République*, ed. Peirre Nora, xvii–xlii. Paris: Gallimard.

Parry, Bronwyn. 2004. Technologies of immortality: The brain on ice. *Studies in History and Philosophy of the Biological and Biomedical Sciences* 35:391–413.

Peeters, Benoît. 2012. *Derrida: A Biography*. Cambridge: Polity.

Radin, Joanna. 2013. Latent life: Concepts and practices of human tissue preservation in the International Biological Program. *Social Studies of Science* 43:484–508.

Serres, Michel. 1982. *The Parasite*. Trans. L. R. Schehr. Baltimore: The Johns Hopkins University Press.

Stoler, Ann Laura. 2009. *Along the Archival Grain: Epistemic Anxieties and Colonial Common Sense*. Princeton: Princeton University Press.

Strathern, Marilyn. 1996. Cutting the network. *Journal of the Royal Anthropological Institute* 2:517–535.

Sunder Rajan, Kaushik. 2006. *Biocapital: The Constitution of Postgenomic Life*. Durham, NC: Duke University Press.

Taussig, Karen-Sue, Klaus Hoeyer, and Stefan Helmreich. 2013. The anthropology of potentiality in biomedicine. *Current Anthropology* 54 (suppl. 7): S3–S14.

Waldby, Catherine, and Robert Mitchell. 2006. *Tissue Economies: Blood, Organs, and Cell Lines in Late Capitalism*. Durham, NC: Duke University Press.

Freezing Species

13 Banking the Forest: Loss, Hope, and Care in Hawaiian Conservation

Thom van Dooren

Introduction

In a single room, tucked away on the main Honolulu campus of the University of Hawai'i, a group of dedicated people have set up an "ark"—a place of last refuge—for some of Hawaii's many highly endangered tree snails.[1] The ark is not a particularly fancy affair: a small room that houses about six "environmental chambers" that look quite a lot like old refrigerators. These units allow staff to control daily temperature and light cycles and have been retrofitted with sprinkler systems to produce a short period of "rain" every eight hours. Inside each environmental chamber are a few small terrariums filled with o'hia and other local vegetation. In this strange environment, some of the rarest gastropods on earth spend their lives.[2]

Among the many endangered snails that I saw during my visit to this ark, one in particular stood out: *Achatinella apexfulva*, a single snail in a terrarium all on its own (figure 13.1)—on its own because this tiny being is now thought to be all that is left of its species. Despite over a decade of searching in the island's forests, scientists have been unable to locate any more. As I stood in that room that day, something about this tiny, solitary figure called out to me, posing new questions about the many problems and promises of contemporary conservation practices. Hope mingles with loss in places like this. We are compelled to hope and care, and yet we must also acknowledge the hopelessness of the situation, the various kinds of losses that are now inevitable.

Of course, tree snails are just some of the many species in trouble in Hawaii's forests. A week earlier I had been on the Big Island, walking through the forest along the upper slopes of Mauna Loa with a group of conservationists, biologists and state and federal land managers. On first

Figure 13.1
The last snail (*Achatinella apexfulva*).

encountering these forests I found it hard to view them as fractured and disappearing landscapes. All around me giant tree trunks rose up to support a lush canopy of leaves and branches. To my eyes this was a magnificent place, filled with an earthy smell and a damp coolness. But as I spent more time exploring these places with knowledgeable people, and reading the relevant literatures, another forest became visible: an historical forest home to a diversity of plants and animals no longer found here. While I have little time for yearnings for "pristine wilderness"—a concept that makes little sense and carries many dangers in Hawaii, as it does in most other places (Cronon 1995, Helmreich 2005)—I remain sympathetic to invocations of a historical forest that aim to draw our attention to key changes and important absences. While the past can never offer a perfect, or even ideal, blueprint for present ecologies (Alagona, Sandlos, and Wiersma 2012; van Dooren 2011), it can certainly call out in a way that attunes us to some of the many forms of life that a place once supported that it no longer does, and perhaps no longer can. Hawaii's forests are a place haunted by such a past.

Beneath the dense rainforest canopy, where there would once have been a lush understory, there is now often little vegetation at all: a result of widespread grazing and rooting by pigs, sheep and other (relatively) newly arrived ungulates. As a result, bare soil is exposed in many places, to wash away with the rains. The understory plants that do survive, and even thrive, in patches of the forest have been identified to me as "noxious weeds"—strawberry guava and various gingers—in places so thick that passage is impossible. These changes have been bad news for many of Hawaii's plant species, which have been grazed or outcompeted onto the endangered species list, and in many cases over the edge of extinction.

Meanwhile, numerous bird and snail species once commonly found in these forests are no longer present. Alongside the general decline of these areas, newly introduced species of various kinds have played a role. For forest birds, predators including cats, mongoose, and rats have taken a huge toll (alongside introduced diseases like avian malaria and toxoplasmosis), while for Hawaii's snails, key threats have been rats, chameleons, and even a predatory snail (Leonard 2008; Cowie, Evenhuis, and Christensen 1995). While recently introduced species often attract the bulk of the attention—easy targets of criticism in Hawaii, as they are elsewhere (van Dooren 2011, 2015)—the decline of the islands' forests is a more complex phenomenon, set within a long history of widespread deforestation for ranching, plantations, and urban development (Duffy 2010; Duffy and Kraus 2008; Lohr 2012; Conant 2012).

All in all, the simple fact is that Hawaii's forests are being radically transformed and biodiversity is being lost at a staggering rate. These forest dynamics are a key part of the reason that Hawaii is now considered to be one of the "extinction capitals" of the world (Groombridge 2008). Even at higher elevations, which are considered to be some of the most healthy forests in the islands, a race is now on to hold back newly introduced species and hold on to some of the "natives" that are disappearing far too quickly. This conservation work includes fencing critical areas wherever possible, poisoning some introduced species while trapping and shooting others, revegetating some areas, and reducing the spread of weeds (Juvik and Juvik 1984; DLNR 2012).

But these ecosystem-level management projects are only one part of the contemporary conservation landscape. This chapter considers the loss of Hawaii's forests, and efforts to conserve them, through focusing on what it might mean to "bank" the forest, or its components. Throughout the islands, and around the world, facilities like the snail ark at the University of Hawai'i are endeavoring to safeguard disappearing species through *ex situ* conservation projects. In contrast to *in situ* forms of conservation that aim to keep organisms, and so their species, living in their "natural habitat" (or wherever it is that now passes for this "mythic" terrain—mythic in the sense of simultaneously imagined and yet very real and often lost), these *ex situ* approaches involve lifting animals or plants out of these environments to be conserved elsewhere, in sites that range from freezers filled with seeds or DNA to zoos and captive breeding facilities.[3]

As more and more of Hawaii's forest species are threatened with extinction, many view these banks as offering a vital site of last refuge. This chapter considers three such banking facilities: starting with a seed bank, and then turning to a bird-breeding center, before finally returning to the tree snail ark. All are fundamentally *hopeful* projects: simultaneously animated by, and working to enable, visions of a future in which the forest—or some semblance of it—might again be possible. Here we see that hope is a central component of the temporal infrastructure of cryopolitics (Radin and Kowal, this vol.), as we strive toward a future that cannot yet be realized but that with effort and care may one day come to pass.

In taking up these topics, this chapter responds to an ongoing call in the environmental movement to focus on hopeful narratives. We are told that "good news stories" instead of "doom and gloom" are what is needed to compel people to appropriate action (Brand 2013; Futerra n.d.; Kelsey and Hanmer 2010). In an article in *Orion* magazine, activist Derrick Jensen has suggested a different tack: he advocates the need to embrace the loss of hope. Hope, he argues, is the longing for something over which we have no agency or control. However, when we take responsibility, when "we realize the degree of agency we actually do have, we no longer have to 'hope' at all. We simply do the work. ... I would say that when hope dies, action begins" (Jensen 2006). I am not satisfied with either of these broad approaches. I don't think there is a simple choice between hope and despair; nor is there a simple contrast between hope and action. Of course, we often don't get to decide in a detached manner how we will respond to the loss of a world; but insofar as we do, rather than advocating one of these generalized affective practices over the other, this chapter argues that what is needed is a critical engagement with the *particular* forms that our hopes for the future take.

In exploring the work of biodiversity banking, this chapter takes up entangled themes of loss, hope and care. What kinds of losses are these banking practices able to stem? How might some hopes work to enable the denial of, or distract from, the urgency of our current period of incredible loss? In short, how is hope for the future of a snail, a bird or a plant cultivated and maintained through situated, hopeful, practices—and at what cost to whom? Drawing on recent work by Eben Kirksey (Kirksey, Shapiro, and Brodine 2013) and Maria Puig de la Bellacasa (2012), this chapter asks what we should be hoping *for* in these times of incredible loss, and are we able to hope responsibly. What might a *careful* form of hope look like—a hope that develops as part of a broader practice of "care for the future"?

Ho'awa: Seeds, Stasis, and the Promise of Vitality to Come

On the northern edge of Honolulu, nestled among forest that stretches back up into the interior of the island of Oahu, is the Lyon Arboretum and Botanical Garden. Part of the University of Hawai'i, it has a multifaceted mandate including research, education, and conservation. A key part of that conservation work—alongside a great deal of research on Hawaiian plants, their status and propagation—is a working seed bank. Filed away in freezers and fridges are around 4 million seeds from 400 different taxa of Hawaiian plants, approximately 168 of which are federally listed as threatened or endangered. At present, roughly a third of Hawaii's 900–1,000 endemic plant species are similarly listed under the Endangered Species Act and as many as 10 percent are already thought to be extinct (Lyon Arboretum 2014a). In this context, this seed bank represents an important conservation strategy. As Hawaii's forests are cleared or transformed, and other ecosystems are similarly degraded, a core part of this bank's mandate is to provide a "backup" for disappearing botanical diversity.

Seed banks like that at the Lyon Arboretum are just one example of the "banking" projects that have become an increasingly important part of contemporary conservation. In addition, a host of other storage options have been, and are being, developed in an effort to hold onto individuals and/or biogenetic material of endangered species. Places like the Frozen Zoo in San Diego bank DNA and gametes, while a range of artificial environments—including captive breeding facilities and zoos—now house the only remaining members of many endangered species (Friese 2013; Chrulew

2011a, this vol.; Kirksey, this vol.). Some of these facilities rely on freezing temperatures that slow biological processes—very literally *cryo* endeavors (the term coming from the Greek κρύο, meaning "icy cold")—while others freeze and preserve in a more metaphorical sense.

Collectively, these are "technologies of stasis," united by a common effort to hold species in limbo: simultaneously inside and out of the world. On one level, they are clearly located *outside* the world of flux and change: this is their explicit purpose, to set aside, preserve, and safeguard. And yet they are also most definitely *in* the world in other ways: it is, after all, the desire to keep species present, to prevent the kind of exit from the world that is extinction, that animates these projects. Not quite *in* the world and yet most definitely *of* the world. Deborah Bird Rose has characterized this space as a "zone of the incomplete" (Rose, this vol.). Organism and samples are not allowed to *live* in any full sense, but nor are they allowed to *die* in any full sense. Instead they are held in a state of "latent life" (Radin 2013). Here, the lives and deaths of individual organisms and of biological samples (be they seeds or DNA) are managed in an effort to ensure the continuity of that most cherished strata of biodiversity, the species.

The seed bank offers an ideal example of this kind of in-between-ness. In the seed bank, storage requires an oscillation between various states, a process called "refreshment." Seeds or vegetative propagation material are stored for a certain amount of time (dependent on temperature and length of viability, which can be markedly different for different species), and then planted out in controlled conditions. Seeds are then selected from these plants and put back into storage.[4] In fact, determining just how long seeds can be stored, and at what temperatures, and still be expected to reliably germinate, is a key part of the work conducted at facilities like the Lyon Arboretum.[5]

Among the many plants banked here are representatives of some of the species known locally as hoʻawa (genus *Pittosporum*). These plants have relatively large and hard capsule-like fruits that contain their seeds. Information on the conservation status of these species is somewhat uncertain, but several of them appear to be rare or declining, with three species now federally listed as endangered (Smithsonian Department of Botany n.d.). One species of hoʻawa—*P. hosmeri*—will enter our story again a little later. While this species does not have an official status on the Endangered Species List, it is today thought to be in decline. As Susan Culliney notes, most of the trees now encountered are older and there is a "general lack of seedlings or saplings in the wild" (Culliney 2011, 21). It seems that in Hawaii's forests, ho'awa has problems with dissemination and germination. At the Lyon Arboretum, however, with a little help from people, these problems have been overcome, at least for the purposes of storage. Here, seeds from ho'awa are cleaned and dried to reduce their moisture content. They are then ready to be packed into airtight containers and frozen for longer-term storage, or refrigerated if they are likely to be used within a few years. As with other seeds stored in this way, samples will be thawed and germinated occasionally to ensure their viability. As the Lyon Arboretum guidelines note: "When the first signs of reduced germination appear, it is time to replenish the seed supply" (Lyon Arboretum 2012).[6] Here, in the space of oscillation between freezer and soil, between seed and plant, what is stored is not simply a sample but an intergenerational lineage of ho'awa, and through it, a species.

But holding in stasis, or something approximating it, is only the first part of the promise of these cryotechnologies. Cryo-accessions are not just "specimens" held for study or display, although they may be this too. Importantly, holdings within cryotechnological spaces are explicitly animated by dreams of (re)vitalization, by visions that they or their progeny might one day be "made vital" (Friese n.d., 2) in some fuller sense through their return to the world beyond.[7] In short, to stretch the metaphor, alongside freezing, these cryotechnologies always also involve practices of *thawing* (Kowal et al. 2013).[8]

It is tempting to draw a clear distinction between cryofacilities involved in endangered species conservation on the one hand, and museums and other such spaces of preservation on the other. While the former stores with the explicit aim of return, the latter seeks to hold on to relics without a similar agenda. However, I am not sure that such a simple line can be drawn. Museum specimens clearly have lives of their own in some remarkable ways, reaching out to teach us about other kinds of worlds and perhaps compel us to enact them (Hatley 2013). Similarly, much of what is held in these cryofacilities will never go anywhere else (except perhaps to another similar facility in the name of maintaining the genetic—or some other form of—diversity that makes the collection self-sustaining and/or valuable). Nonetheless, there is something important—which cannot be lost sight of—in the fact that these cryotechnology facilities are in large part motivated and animated by hope for a future in which these species are restored to the wider world.

Here we see the generative nature of freezing, whether literal or metaphorical. The "holding still" that is achieved through freezing is itself a powerful enabler of movement (Radin and Kowal, this vol.). As Michael Bravo has argued, the stasis that freezing enables, the space simultaneously inside and out of the world that it creates, allows frozen entities to travel through space and time in remarkable ways (Bravo, this vol.). In the case of endangered species, it is primarily this movement through time that animates cryotechnological hopes: it is the capacity of that which is frozen to survive into a potentially better time when life in the wider world might again be possible. According to the Lyon Arboretum's website, the seed bank collection is "akin to a Noah's Ark or insurance policy for endangered plants" (2014b). It is a safe place to weather the Anthropocene storm.

And so, while these banking facilities are sites underwritten by intense processes of loss, they are also important sites for the production and maintenance of hope, containing within them the possibility that at some time in the future, after the wreckage has cleared, at least a semblance of the forest might again be possible. I view the Lyon Arboretum seed bank as an example of what Eben Kirksey has called "modest forms of biocultural hope" (Kirksey, Shapiro, and Brodine 2013). Kirksey is writing with and against Jacques Derrida, and in particular the emptiness, or indeterminacy, of Derrida's notion of the "to come," of a messianicity without messianism (Derrida and Ferraris 2001). Here, Derrida emphasizes a relationship of radical openness toward the future that is not locked down to any particular vision or project. Grounded in the notion of the event, the unexpected, that ruptures temporal continuities in the name of something wholly new, Derrida sees a primary responsibility in remaining open to the unpredictable, the incalculable.[9]

In contrast, Kirksey emphasizes the need for more *grounded* hopeful projects, engaged in practical and concrete acts of care for the ongoing biological and cultural richness of our world. These are not utopian visions that hope to set everything to rights in one fell swoop, but modest efforts to make a difference in often creative and inclusive ways that draw others into an opening, rather than recruiting them into a fixed vision of how things might be. While in some ways the seed bank is a project that locks in a strategy and aims for a fixed future, in other ways it is grounded in precisely this kind of openness. For example, alongside the more formal seed bank, this facility plays a key role in helping to keep Hawaiian seeds circulating in the wider world, allowing community members to bank their own seeds until they are needed for restoration projects, and also providing ongoing research and instruction to the wider community about how they might bank and germinate seeds at home (Yoshinaga and Kroessig 2012; Lyon Arboretum 2012).

David Wood (2006) makes an important observation about these kinds of hopeful projects. Again, thinking with but also against Derrida, Wood advocates for the importance of forms of hope that aren't so much grounded in expecting the unexpected, but rather in responding in practical and meaningful ways to a widespread problem in contemporary society of "unexpecting the expected"—that is, burying one's head in the sand and refusing to face up to the reality of what is almost certain to happen (i.e., what should reasonably be expected). Like Kirksey's modest projects, Wood is calling for "grounded" forms of hope—grounded metaphorically, but perhaps also literally. These are hopeful projects that get messy and muddy, down in the soil, responding where we can to challenges we can already see around us in an effort to build a better future, however partial and uncertain our vision of it may be. This seed bank is, in many ways, just such a project.

But for all the possibilities that projects like this one open up, they also come with a range of limitations and dangers that require attention. Ho'awa provides us with a way into these issues. Moving from the security of the bank, back to the forest, we might, for example, ask after the birds who once ate the big orange fruits of the ho'awa plant.

'Alalā: Birds and Frayed Relationalities

I vividly remember my first encounter with 'alalā (the Hawaiian crow, *Corvus hawaiiensis*). As I slowly stepped into the long thin aviary, five young birds turned to look at me. Almost immediately they began to move silently in my direction; wings outstretched, half jumping, half floating, they gracefully moved between the upper branches of the aviary. Clearly intelligent and inquisitive, they assembled above me and began to caw loudly, watching me closely. This encounter took place at the Keauhou Bird Conservation Center (KBCC), a small collection of aviaries near the top of Kilauea on

Hawaii's Big Island that house some of rarest birds on earth (figure 13.2). Here, among a variety of smaller honey eaters—like the Palila and the Maui Parrotbill—are around sixty 'alalā. Now extinct in the wild, these birds—along with an even smaller group at a sister facility on Maui—are all that remains of their species. In addition, one reproductively "uncooperative" bird now lives at the San Diego Zoo, home of the Frozen Zoo, where staff are attempting to obtain semen samples that might be put back into the captive breeding program.[10]

The decline of the 'alalā is a familiar story. As fruit and flower specialists these birds were tied to the island's changing forests. Over the past couple of centuries in particular, as these places became increasingly degraded, and as new predators and avian diseases arrived, 'alalā began to slip out of the world. Importantly for this story, 'alalā are also thought to be the last remaining birds that might effectively disperse seeds from ho'awa (*P. hosmeri*, and perhaps other related species). Although Hawaii's forests were once home to many large birds, including large flightless birds, almost all of them are now gone. Of the 113 bird species known to have lived exclusively on the Hawaiian islands just prior to human arrival, almost two-thirds are now extinct. Of the 42 species that remain, 31 are federally listed under the Endangered Species Act (Leonard 2008). Undoubtedly, some of these birds would have eaten the large fruits of ho'awa to transport their seeds around the islands. Beyond simply moving seeds, these birds also likely played an important role in helping them to germinate. It seems that without having traveled through a birds' digestive system—or at the very least having had the fruit stripped away—ho'awa seeds are unable to germinate. But now only 'alalā remain to take up this role. As the most recent study put it, "Our results indicate that ho'awa relies entirely on 'Alalā ingestion or manipulation for germination" (Culliney et al. 2012, 1727). In this context it has been suggested that ho'awa may now be an "ecological anachronism" (Culliney et al. 2012, 1729): a species adapted to a way of life, in this case an interspecies relationship, that is no longer present (Janzen and Martin 1982; Barlow 2000).

Taken together, 'alalā and ho'awa remind us that, as Maria Puig de la Bellacasa (2012) has succinctly put it (drawing on Donna Haraway), "nothing comes without its world." Long histories of coevolution, of entangled becoming, have placed these species at stake in each other. At their extreme, these relationships of dependence can give rise to coextinction: without a

Figure 13.2
A captive 'alalā (*Corvus hawaiiensis*).

vital pollinator or disperser, a plant species might wither and fade until it disappears altogether. Or, the loss of a disperser may restrict a plant species' range and distribution in such a way that it becomes increasingly vulnerable to disease or other stochastic events, and less and less adaptable to climatic and other variability.

Here, we see that banking living 'alalā in captive aviaries, or their gametes or DNA in a frozen zoo, will not help ho'awa and those other plants tangled up with these birds in co-constitutive relations of dependency. In an important sense, for ho'awa and the broader forests of Hawaii, 'alalā is already extinct. The world is already unraveling.

All cryotechnologies of endangered species conservation have in common the fact that, to a greater or lesser extent, they fail to *conserve* what many living beings want to hold on to. There is a kind of reductivism underlying these *ex situ* approaches in which "accessions"—be they living birds, seeds, or DNA samples—must be held outside of the vast webs of relationship that have given rise to and sustained them. These relationships are variously glossed as "evolutionary," "ecological," or "cultural," but they are always more complex and entangled than these labels imply.

In the case of 'alalā, ho'awa and a range of other plants are now placed at increased risk of extinction. More broadly, as biologist Paul Banko has noted, the absence of this last large seed disperser will "potentially influence the composition and function of dry- and wet-forest ecosystems" in Hawaii (Banko, Ball, and Banko 2002). In this way, a broad range of other species will likely be drawn into the changing forests taking form around this absent crow. Among them will be humans: for example, the Hawaiians who collect and utilize these plants for traditional cultural practices (Culliney 2011), and those for whom 'alalā is itself an aumakua or ancestral deity (US Fish and Wildlife Service 2009). These complex relationships are fraying and unraveling; relationships that have been or will be lost even while 'alalā, as a species, lives on in secure aviaries. If 'alalā makes it back to the island's forests one day, some of these relationships might be restored—for others it may well be too late.

Here we see some of the risks associated with this form of conservation-as-banking. As Matthew Chrulew has noted, these cryoprojects are grounded in an effort to "secure life against *living itself*," to "protect life from much that essentially characterizes living: its embodiment, its relationships to the world, to the environment, to ancestors and offspring, to other creatures, to human beings, to the passage of time, to reproduction and intergenerational transmission," and the list could go on (Chrulew, this vol.). While seed from ho'awa (and a whole host of other species at risk) might itself be tucked away in freezers, the forest is a complex biosocial achievement from which individual species or even groups of plants and animals cannot simply be banked, thawed out later, and then slotted back in. The multiple interwoven and shifting relationships that constitute the forest—like any other ecosystem—cannot be put on ice.

In this context, it is clear that extinction is not the death of the last of a species, as is so often assumed. Rather, it is a slow unraveling of delicately interwoven *ways of life*. The extinction of 'alalā *is* the decline of hoʻawa, and a range of other subtle shifts in forest composition; it *is* the potential loss or transformation of Hawaiian stories and cultural practices; it *is* an altered experiential landscape such that, as one local put it, "the Hawaiian forest is now missing its most intelligent and charismatic" presence (van Dooren 2014b). These are the unravelings that constitute extinction, often beginning long before the death of the last individual of a kind and rippling out into the future long afterward. This is an entangled and ongoing process of loss, what I have elsewhere called "the dull edge of extinction" (van Dooren 2014a).

But the losses taking place despite the banking of 'alalā go deeper than this. In an important sense, like the forest, this species cannot be banked. A central part of what is at issue here when it comes to the *ex situ* conservation of animals is now often glossed as "behavior." But behavior is far too miserly a term. Depending on the species in question, a whole range of learned behaviors, vocal repertoires, and social skills—what some have referred to as animal "cultures" (Lestel 2002)—often require processes of interaction and learning that are not, and in some cases cannot be, conserved *ex situ*. If released, will these captive bred animals act and live as their forebears did—will they possess the necessary behaviors to survive (van Dooren 2016)? Many of these behaviors are of central concern in the conservation of 'alalā, and efforts are being made to retain as much of this behavioral diversity as possible—and even to teach them lost skills. For example, it is feared that in captivity these birds may have lost a great deal of their vocal repertoire. Despite effort and significant achievements, Alan Lieberman (the former director of this conservation program) notes that "We haven't been able to transfer the 'alalā culture from one generation to the next. But we'll do the best that we can and we'll create a new culture."[11]

Things can get even more complex when animal species are banked in the form of frozen gametes or DNA that might one day allow extinct species to be resurrected with the help of biotechnologies and animal surrogates (Friese 2009, 2013; Chrulew 2011b). That these projects are called "conservation" at all is itself deeply problematic. As has been noted by others, the notion that a species can be meaningfully "captured" in these ways recapitulates a highly reductive and simplistic notion of DNA as blueprint (Thacker 2003, 73). For many animal species, especially highly intelligent

social learners like 'alalā, a great deal of what makes the species who it is—
its evolved and evolving "way of life"—as well as much of what enables
the species to survive in the wider world, simply cannot be conserved in
this way.

In exploring these issues we are alerted to the fact that there are important distinctions between various kinds of banking projects. Different ways
of storing species make a difference. In addition, the only way to keep many
relationships alive and well is through *in situ* conservation, or at the very
least for the bank to play a dual role: safeguarding but also rearing and
releasing. Both the seeds in the Lyon Arboretum and the birds at KBCC are
part of such efforts. Both facilities are actively involved in ongoing projects
to keep many of their endangered charges alive in the wider world, now or
in the near future. In these and other ways, these facilities work to acknowledge their own limitations. They acknowledge that they are far from ideal,
that they are approaches of last resort. For now, however, the simple fact
remains that without KBCC, 'alalā would most definitely be irrecoverably
extinct. However difficult its return to the forest may be, however imperfect
its current situation, perhaps we can take hope from the fact that some kind
of survival is still possible.

Kāhuli: Snails, Hope, and Loss

In the Hawaiian islands, where endangered species status is commonplace,
the precarious future of many taxa of tree snails still manages to stand out
as particularly worrying. In all, an estimated 75 percent of the more that
700 named species are thought to already have been lost (Cowie, Evenhuis,
and Christensen 1995). The snail ark at the University of Hawai'i mentioned at the outset of this chapter focuses on a particular genus, *Achatinella*
(known locally as Kāhuli). Of the forty-three species of this genus once
found on the island of Oahu, only ten remain—all of which are federally
listed as endangered. In this dire context, Mike Hadfield, the founder of this
ark, explained the project to me as "a last-ditch effort to 'save snails' that
would certainly have been devoured by alien predators ... in the immediate
future."[12] In short, it was a response to a crisis, an effort to make possible at
least some kind of future for these species. But in what sense have their
extinctions really been delayed in this ark? What kinds of futures, what
kinds of hopes, should we really entertain on their behalf?

The sole remaining individual of *Achatinella apexfulva*—the single snail in a terrarium all on its own (figure 13.1)—offers a paradigmatic example of the complexity of this space. This snail is a tragic exemplar of the "nonpossibility" of at least some banked life. When it dies, a whole evolutionary lineage will pass from the world, and yet at the same time nothing much will change. This species is already among what some biologists call the "living dead": it is a species whose population has become so small that extinction in the near future is now inevitable (Sodikoff 2013).

Alongside this obvious example, many of the other snail populations in the ark are also in dire trouble. In Hadfield's words, "after functioning very well for more than fifteen years, something changed a few years ago, and most of the lab-snail populations have gone into severe decline." Inbreeding within small isolated populations seems to be at least partly to blame. Wherever possible staff are working to introduce new snails into these populations to increase "genetic vigor." But for some of the species in the ark, there are now no—or very few—other survivors to draw on. And so, for at least some of these species, there will likely be no release. The space of latency created by the bank is not one of potential life, but rather only of incomplete death where there is no hope of animating desired futures (Rose, this vol.; Kowal and Radin 2015). The "ark" is in reality something more like a living tomb.

In fact, all of Hawaii's endangered tree snails find themselves in a pretty bad situation. Restoring habitat for these species will be an incredibly difficult task, if it is possible at all. As mentioned earlier, the primary threat to these species is a diverse group of introduced predators—rats, chameleons, and a carnivorous snail. Even the ants introduced have been known to kill local snails. All of these species are very difficult to control in a forest environment. In relatively small spaces, conservationists have worked tirelessly to establish habitat for the release of captive bred snails. With assistance and funding from the US Army they have set up high-tech barriers incorporating electric fences, video surveillance, and a range of other devices to both exclude predators and monitor the barrier. It is then a matter of eradicating all predators inside the fenced area along with ongoing vigilance to ensure that they don't establish themselves again (Hadfield 2014, pers. comm.). This approach can only be applied on a small scale and has a questionable chance of long-term success. For now, however, it may see these snails through a little longer.

In the tiny bodies of these gastropods, we see that, in many cases, what these banks actually freeze and delay is not so much extinction itself, but rather the *recognition* of extinction, the recognition that something significant has already been lost. Single individuals or declining populations stand in for this thing called a species, keeping it off the official listings of the departed. This way, when that moment of recognition does finally arrive—a moment that is still often far too closely tied to the death of the last individual—extinction has been so long coming that no one can really be surprised. After so long in cold storage, even the long drawn-out ripples of loss and change that constitute the dull edge of this extinction will have largely settled into new patterns of life and death. Here we encounter the dangerous side of at least some practices of hope, of working to imagine a better future. If it is a future that cannot come—if it is a vision grounded in seeing latent life where there is only living death (at least for the larger species)—then this particular hope is no longer helpful.

Hope is often associated with the affirmation of life, the refusal to give up, and consequently the absence of hope is associated with despair. As Mary Zournazi puts it: "Without hope what is left is death—the death of spirit, the death of life—where there is no longer any sense of regeneration and renewal" (2002, 16). But sometimes affirming life is not what is needed. Instead, hope for ongoing life becomes a form of denial that allows us to go on without having to come to terms with our reality or with the vital need for change. In this way, these hopeful banking projects enable the laundering of what biologists call our "extinction debt" (Sodikoff 2013)—rendering invisible all those extinctions that are now inevitable as a result of past actions, extinctions that are already unraveling the world in various ways. In so doing, these banks, whatever their intentions, play an important role in undermining our imaginative and moral capacity to perceive the pressing crisis of the current mass extinction event.

Beyond delaying the recognition of extinction, these banks also have the potential to delay much-needed conservation action. In some cases they might have the opposite effect, providing strong incentives to deal with larger conservation issues to create the habitat necessary for a population's release; in fact, the unexpected success of the ʻalalā captive breeding program is providing precisely this pressure as they run out of space to house birds. But, these populations can equally be an excuse to delay that action further—resting in the comfort that we have a secure backup.

Cryotechnologies can be mobilized in either rhetorical direction. The more readily and reliably their accessions can be stored away for another day—a seed as compared to a fleshy crow—the more likely they are to become tomorrow's problem. A worrying analogy can be drawn here with debates about ecological restoration and the way in which the rhetoric and quasi possibility of "putting things back later" has been captured by mining companies and others interested in the continuation of business as usual. If species, or perhaps even whole ecosystems, can be banked and thawed out later—or rather, if the perception can be created that they can be—then new possibilities open up for exploitative practices in fragile places. As Chantal Mouffe reminds us, "Hope can be something that is played in many dangerous ways" (2002, 126).[13]

This surely is not the intention of the many committed individuals who dedicate so much of their lives to these and other biodiversity banking initiatives, but in dark times the lure of hope as a form of denial or distraction can be very strong. There is a strange similarity, although far from an equivalence, between Derrida's appeal for a radical openness toward the future, a hopeful invocation of the "to come," and the vague way in which environmentalists now often urge one another to focus on the positive. In this context, what is hoped *for* often seems less important than the act of being hopeful, of encouraging others into a particular state of being toward the future. But vague and general "hope" is not always helpful. Kirksey reminds us of the importance of specific hopeful projects that are grounded and partial, but still open. As part of this effort, what is needed is a critical lens on, and more attention toward, what it is that we are specifically hoping and working toward. As Ghassan Hage has argued, "We need to look at what kind of hope a society encourages rather than simply whether it gives people hope or not" (2002, 152). What should we be hoping *for* in these times of incredible loss? Are we able to hope responsibly, perhaps carefully? Which is to say, can our hopes be translated into meaningful action and taken up in a way that recognizes the myriad losses and exposes the dangers that lie buried in the things we hope might yet come to pass?

In this context, we might view responsible hope as a practice of "care for the future." Care must be understood here as something far more than abstract well-wishing. As Puig de la Bellacasa has noted, a thick notion of care requires that it be understood as simultaneously, "a vital affective state, an ethical obligation and a practical labour" (2012, 197). To care for another, to care for a possible world, is to become emotionally and ethically entangled and consequently to get involved in whatever practical ways we can. But, as Haraway notes, caring deeply also "means becoming subject to the unsettling obligation of curiosity, which requires knowing more at the end of the day than at the beginning" (2008, 36). Knowing more, in this sense, is about being drawn into a deep contextual and critical knowledge about the object of our care: *what* am I really caring for, *why*, and at *what cost* to whom? The grounded and responsible hope that we need today—hope for a world still rich in biocultural diversities of all kinds—requires affective, ethical, and practical action, but it also requires a critical engagement with the means and consequences of its own production.

Ultimately, I don't think this means we should abandon our banking projects. Importantly, some of them will "work"—in the sense of holding on to species and even getting them back out in the world in meaningful ways. Even where this will not be possible, I understand the desire to hold on to individuals like the last *Achatinella apexfulva* for as long as we can—provided that they are living flourishing lives—as an effort to cultivate some semblance of responsibility for another whose world we (collectively) have destroyed. Even where there is no hope for recovery, perhaps we are still called into the effort to achieve some sort of grace, "some healing of our condition on this earth and thereby, with it, some measure of cosmic repair" (Levene 2013, 164).

For all of these reasons—both practical and ethical—my intention in this chapter has not been to argue for the abandonment of banking projects. Rather, my goal has been to unsettle the notion that banking and embracing hopeful futures are unproblematically good. Instead of abandoning these banking projects, my hope is that we might reimagine and rework them in ways that would enable us to develop fuller forms of responsibility for all those things that we cannot quite hold on to and all those that we cannot ever restore. This kind of bank would be one in which its own limitations—the compromises and challenges both within its walls and in its relations with a wider world—were made *visible* and made a site for the ongoing *reworking*, the improvement, of our relations with disappearing others. In short, it would be a place that continually asks of itself and its practices, why and at what cost to whom (van Dooren 2014a, 116–122). In this way, banking might become more than a practice of setting aside for the future; it would also be an active site of intervention into the present.[14]

While in times like these we certainly need all the conservation efforts that we can muster, it remains vital that we pay careful attention to the means by which particular approaches generate and sustain their visions for the future. For at least some species, the time for hopefulness has passed. It is time for us to acknowledge, to take responsibility and care for, other kinds of futures.

Notes

1. The research that informs this chapter was funded by the Australian Research Council (DP110102886).

2. I am grateful to Professor Mike Hadfield, the founder of this facility, for offering to give me a tour (in early 2013) and for subsequently being willing to discuss this conservation work by email. However, the views expressed in this chapter are my own.

3. In reality, however, things are far more complex than this simple distinction implies. Part of the difficulty lies in determining just where a plant or animal "ought" to be: what counts as "natural habitat," a line of questioning made all the more difficult in an age of movement initiated by climate change and large-scale environmental transformations of various kinds. Furthermore, as Whatmore and Thorne (2000) have noted, the prospect of a simple line between *in situ* and *ex situ* is grounded in a series of problematic dualistic frameworks that position humans outside of "natural" systems while simultaneously failing to account for all of the messy ways in which actual conservation practices redo the world in less monochromatic hues. In this context, it is often difficult to know what to make of this division. There is undoubtedly something in the difference between these spaces and approaches, but it is not as simple as references to "natural habitat" would have us believe.

4. While this process of refreshment in some ways epitomizes the kinds of in-between-ness practiced in banking projects, Emma Kowal has noted that from another perspective the field and the freezer might be viewed as complementary parts of the same in-between state of latency (pers. comm., March 2014).

5. Global catalogs, like the Kew Seed Information Database, compile and share this information with researchers and practitioners around the world. See http://data.kew.org/sid/.

6. All of this is relatively straightforward for ho'awa (which is classed as an "orthodox" seed), but for some other plants—usually referred to as "recalcitrant," refusing

to cooperate in their own conservation—special conditions sometimes need to be introduced, such as avoiding freezing or engaging in more frequent replenishment (Lyon Arboretum 2012).

7. Or, at the very least, that they might contribute to research—for example reproductive or genetic research—toward this end.

8. As Joanna Radin and Emma Kowal (2015) have noted, all frozen objects have a potentially thawed future that may be accidental or anticipated, feared or eagerly awaited. In this context, thawing—or revitalization—is always also a process of what Radin and Kowal have called "resocialization"—a movement in which the frozen entity needs to somehow be "resynced" with a wider world that will likely have gone on changing in various ways. This resyncing is a topic that I will return to below.

9. I am not sure that Derrida's approach is as empty of content as it first appears. He is centrally occupied with bringing about a better future. In this context, the "to come" takes the form of what Paul Patton has called "a promise": "it is a means by which an imagined future can intervene in or act upon the present. Just as a promise in relation to some future state of affairs has consequences for one's actions in the present, so the appeal to justice or to a democracy to come will have consequences in the present" (Patton 2004). Here, Derrida's broader body of work gives many indications of the kinds of "projects"—democracy, justice, hospitality—that might animate and guide our actions in the present. The fact that justice, for example, will never "arrive" in the present—that perfect justice is *impossible*—is precisely what gives it, and will continue to give it, the capacity to motivate *better* futures (Derrida and Ferraris 2001; Mouffe and Laclau 2002, 128).

10. Alan Lieberman, interview with the author. All references to Alan Lieberman are from an interview conducted by the author with Lieberman, then Director of Regional Conservation Programs at the Institute for Conservation Research, San Diego Zoo, on December 1, 2010.

11. For a detailed discussion of how behavior is understood, managed, and conserved in the 'alalā project see van Dooren 2016.

12. All references to Mike Hadfield refer to a discussion during my visit to the University of Hawai'i in January 2013 or a personal email correspondence with him in February 2014.

13. This is a pattern of response that is evident across many domains of freezing technology, where freezing life emerges as a mode of "problem solving" that fails to get at the roots of a larger problem, and may in fact exacerbate these issues. For example, Inhorn notes that the freezing of women's eggs is often presented as a mode of enabling women to delay childbirth, to "have it all." In reality, however, this is a solution that fails to grapple with the larger structural, social, economic, and

political issues that often prevent women from pursuing a career *while* they have children (Inhorn, this vol.).

14. Here, my position links back up with Derrida's argument in an interesting way. The bank becomes a kind of "promise," in Paul Patton's sense of the term (see endnote 9).

References

Alagona, Peter S., John Sandlos, and Yolanda F. Wiersma. 2012. Past imperfect: Using historical ecology and baseline data for conservation and restoration projects in North America. *Environmental Philosophy* 9 (1): 49–70.

Banko, Paul C., Donna L. Ball, and Winston E. Banko. 2002. Hawaiian crow (*Corvus hawaiiensis*). In *The Birds of North America Online*, ed. A. Poole. Ithaca: Cornell Lab of Ornithology.

Barlow, Connie. 2000. *The Ghosts of Evolution: Nonsensical Fruit, Missing Partners, and Other Ecological Anachronisms*. New York: Basic Books.

Brand, Stewart. 2013. The dawn of de-extinction: Are you ready? http://longnow. org/revive/de-extinction/2013/stewart-brand-the-dawn-of-de-extinction-are-you -ready (accessed May 19, 2014).

Chrulew, Matthew. 2011a. Managing love and death at the zoo: The biopolitics of endangered species preservation. *Australian Humanities Review* 50:137–157.

Chrulew, Matthew. 2011b. Reversing extinction: Restoration and resurrection in the Pleistocene rewilding projects. *Humanimalia* 2 (2): 4–27.

Conant, Sheila. 2012. Fifty years of SPAM: Reflections on conservation in Hawaii. Paper presented at the Hawaii Conservation Conference, Honolulu, July 31–August 2.

Cowie, Robert H., Neal L. Evenhuis, and Carl C. Christensen. 1995. *Catalog of the Native Land and Freshwater Molluscs of the Hawaiian Islands*. Leiden, the Netherlands: Backhuys Publishers.

Cronon, William. 1995. The trouble with wilderness: Or, Getting back to the wrong nature. In *Uncommon Ground: Toward Reinventing Nature*, ed. William Cronon. New York: W. W. Norton.

Culliney, Susan Moana. 2011. Seed dispersal by the critically endangered Alala (*Corvus hawaiiensis*) and integrating community values into Alala (*Corvus hawaiiensis*) recovery. Master's thesis, Department of Fish, Wildlife and Conservation Biology, Colorado State University.

Culliney, Susan, Liba Pejchar, Richard Switzer, and Viviana Ruiz-Guitierrez. 2012. Seed dispersal by a captive corvid: The role of the 'Alala (*Corvus hawaiiensis*) in shaping Hawai'i's plant communities. *Ecological Applications* 22 (6): 1718–1732.

Derrida, Jacques, and Maurizio Ferraris. 2001. *A Taste for the Secret*. Cambridge: Polity Press.

Department of Land and Natural Resources (DLNR). 2012. *Ka'u Forest Reserve Management Plan*. Honolulu: Department of Land and Natural Resources, Division of Forestry and Wildlife, State of Hawai'i.

Duffy, D. C., and F. Kraus. 2008. Taking Medawar's medicine: Science as the art of the soluble for Hawaii's terrestrial extinction crisis. *Pacific Conservation Biology* 14:80–88.

Duffy, Deidre. 2010. An historical analysis of hunting in Hawaii. Master's thesis, Natural Resources and Environmental Management, University of Hawaii.

Friese, Carrie. 2009. Models of cloning, models for the zoo: Rethinking the sociological significance of cloned animals. *Biosocieties* 4:367–390.

Friese, Carrie. 2013. *Cloning Wild Life: Zoos, Captivity, and the Future of Endangered Animals*. New York: NYU Press.

Friese, Carrie. n.d. The reproductive logics of cloning: Transforming time, space, and bodies in the biopolitical apparatuses of endangered species conservation. Unpublished paper.

Futerra. n.d. *Branding Biodiversity: The New Nature Message*. Futerra Sustainability Communications.</other>

Groombridge, Jim. 2008. Hawaii: Extinction capital of the world. http://planetearth. nerc.ac.uk/features/story.aspx?id=129 (accessed May 19, 2014).

Hage, Ghassan. 2002. "On the side of life"—joy and the capacity of being. In *Hope: New Philosophies for Change*, ed. Mary Zournarzi. Annandale, NSW: Pluto Press.

Haraway, Donna. 2008. *When Species Meet*. Minneapolis: University of Minnesota Press.

Hatley, James. 2013. Dating extinction: Why and why not specimens matter. Paper presented at Australian Animal Studies Group Conference, University of Sydney, July 7–10.

Helmreich, Stefan. 2005. How scientists think; about "natives," for example: A problem of taxonomy among biologists of alien species in Hawaii. *Journal of the Royal Anthropological Institute* 11 (1): 107–128.

Janzen, Daniel H., and Paul S. Martin. 1982. Neotropical anachronisms: The fruits the Gomphotheres ate. *Science* 215 (4528): 19–27.

Jensen, Derrick. 2006. Beyond hope. *Orion* (May–June): 14–17.

Juvik, J. O., and S. P. Juvik. 1984. Mauna Kea and the myth of multiple use endangered species and mountain management in Hawaii. *Mountain Research and Development* 4 (3): 191–202.

Kelsey, Elin, and Clayton Hanmer. 2010. *Not Your Typical Book about the Environment.* Toronto: Owlkids.

Kirksey, S. Eben, Nick Shapiro, and Maria Brodine. 2013. Hope in blasted landscapes. *Social Sciences Information* 52 (2): 228–256.

Kowal, E., and J. Radin. 2015. Indigenous biospecimens and the cryopolitics of frozen life. *Journal of Sociology* 51 (1): 63–80.

Kowal, Emma, Joanna Radin, and Jenny Reardon. 2013. Indigenous body parts, mutating temporalities, and the half-lives of postcolonial technoscience. *Social Studies of Science* 43 (4): 465–483.

Leonard, David L., Jr. 2008. Recovery expenditures for birds listed under the US Endangered Species Act: The disparity between mainland and Hawaiian taxa. *Biological Conservation* 141:2054–2061.

Lestel, Dominique. 2002. The biosemiotics and phylogenesis of culture. *Social Sciences Information* 41 (1): 35–68.

Levene, Mark. 2013. Climate blues, or How awareness of the human end might reinstil ethical purpose to the writing of history. *Environmental Humanities* 2:147–167.

Lohr, Cheryl Anne. 2012. Human dimensions of introduced terrestrial vertebrates in the Hawaiian Islands. PhD dissertation, Natural Resources and Environmental Management, University of Hawai'i.

Lyon Arboretum. 2012. *Seed Bank Users Guide.* Lyon Arboretum.

Lyon Arboretum. 2014a. *Hawaiian Rare Plant Program.* https://manoa.hawaii.edu/lyonarboretum/research/conservation/hrpp/ (accessed September 7, 2016).

Lyon Arboretum. 2014b. *Seed Conservation Laboratory.* https://manoa.hawaii.edu/lyonarboretum/research/conservation/hrpp/seed-conservation-laboratory-hawaiian-rare-plant-program/ (accessed September 7, 2016).

Mouffe, Chantal. 2002. Hope, passion, politics. In *Hope: New Philosophies for Change,* ed. Mary Zournarzi. Annandale, NSW: Pluto Press.

Patton, Paul. 2004. Politics. In *Understanding Derrida,* ed. Jonathan Roffe and Jack Reynolds. London: Continuum.

Puig de la Bellacasa, Maria. 2012. "Nothing comes without its world": Thinking with care. *Sociological Review* 60 (2): 197–216.

Radin, Joanna. 2013. Latent life: Concepts and practices of human tissue preservation in the International Biological Program. *Social Studies of Science* 43 (4): 484–508.

Smithsonian Department of Botany. n.d. Flora of the Hawaiian Islands. Smithsonian National Museum of Natural History. http://botany.si.edu/pacificislandbiodiversity/hawaiianflora/result.cfm?commonyes=x&genus=pittosporum (accessed May 19, 2014).

Sodikoff, Genese Marie. 2013. The time of living dead species: Extinction debt and futurity in Madagascar. In *Debt: Ethics, the Environment, and the Economy*, ed. Peter Y. Paik and Merry Wiesner-Hanks. Bloomington: Indiana University Press.

Thacker, Eugene. 2003. What is biomedia? *Configurations* 11:47–79.

US Fish and Wildlife Service (UFWS). 2009. Revised recovery plan for the 'Alalā (*Corvus hawaiiensis*). Portland, OR: US Fish and Wildlife Service.

van Dooren, Thom. 2011. Invasive species in penguin worlds: An ethical taxonomy of killing for conservation. *Conservation and Society* 9 (4): 286–298.

van Dooren, Thom. 2014a. *Flight Ways: Life and Loss at the Edge of Extinction*. New York: Columbia University Press.

van Dooren, Thom. 2014b. Life at the edge of extinction: Spectral crows, haunted landscapes, and the environmental humanities. *Humanities Australia* 5.

van Dooren, Thom. 2015. A day with crows: Rarity, nativity, and the violent-care of conservation. *Animal Studies Journal* 4 (2).

van Dooren, Thom. 2016. Authentic crows: Identity, captivity, and emergent forms of life. *Theory, Culture & Society* 33 (2): 29–52.

Whatmore, Sarah, and Lorraine Thorne. 2000. Elephants on the move: Spatial formations of wildlife exchange. *Environment and Planning. D, Society & Space* 18:185–203.

Wood, David. 2006. On being haunted by the future. *Research in Phenomenology* 36 (1): 274–298.

Yoshinaga, A., and Timothy Kroessig. 2012. Guidelines for successful seed storage. Honolulu: Center for Conservation Research and Training, Lyon Arboretum.

Zournarzi, Mary. 2002. *Hope: New Philosophies for Change*. Annandale, NSW: Pluto Press.

14 Freezing the Ark: The Cryopolitics of Endangered Species Preservation

Matthew Chrulew

Heini Hediger's *Wild Animals in Captivity*, published in German in 1942 and in English translation in 1950, became something of a bible among twentieth-century zoo directors for its advice on techniques for producing "natural" behavior and reproduction in captive animals. In this zoo biology manual, temperature was thematized as a variable in the quality of space. Among other elements, such as shade and sunlight, humidity, shelter, and ventilation, temperature was understood as an aspect of animal enclosures that could be modified to produce "the artificial, optimum climate"—that is, a "man-made" environment that mimicked or translated the best natural conditions for the species (Hediger 1964, 79). "Every species of animal prefers a definite temperature, the favourite temperature," wrote Hediger, "and there is reason to think that this is also the most suitable temperature for the species *i.e.* the optimum temperature" (80). While he maintained, following the experiments of animal trader and zoo innovator Carl Hagenbeck, that animals' temperature requirements were often overestimated—for example, tropical species were capable of thriving in European winters—nonetheless, the inevitable restrictions of captive exhibits made it necessary to provide them with a range of microclimates for self-regulation. Temperature was a relative element of the environment that "man" could experimentally determine and reproduce. Hediger related the statistics for a number of species—varying according to their warmer or colder "biotope" and "manner of life"—and gave examples where getting the temperature right led to healthy weight gain and other signs of biological flourishing (80). Providing an optimum rather than a minimum temperature allowed captive animals to live well rather than merely survive.

Such artificial optimization of temperature has only been refined by zoo biologists in the ensuing half century (Hosey, Melfi, and Pankhurst 2009,

319); at the same time, much else has changed. A new focus on conservation has modified zoos' operations; biotechnologies have enabled novel
interventions into their wards' reproductive lives; the rise of genetics has
emphasized the genome informing the organism and population; and continued global development has further diminished native habitats. In particular, the awareness of a changing climate has undermined the stability of
the natural biotope as a referent. Between long frozen mammoths exposed
in melting tundra and threatened polar bears marooned on melting ice
caps, climate change reveals the extinctions of the past as it produces those
of the present. In an epoch when temperature, among other elements of
natural environments, is subject to unpredictable anthropogenic change,
and the threat of extinctions is multiplied by this ecological instability,
zoo practices for guiding adaptation to unfamiliar climates have become
increasingly generalizable to wider domains of wildlife management, and
zoos themselves look to new techniques to ensure the perpetuation of
declining species. In addition to modulating the optimum temperature in
the organism's relationship to its milieu, zoo-based laboratories have begun
to manipulate extreme temperatures in order to hold the species itself in
suspension.

The Frozen Ark Project

In the last few decades, cryopreservation has emerged as a prominent strategy for endangered species protection. With the advancement of freezing
technologies and the rise of genetics and molecular biology, "suspended *ex
situ*" (Soulé 1991, 748) management programs best known as "frozen zoos"
have been proposed as insurance policies, stockpiling invaluable genetic
information against the loss of biodiversity. While various solitary collections have long been established, most prominently at the San Diego
Zoo (which has trademarked the term "Frozen Zoo"), the Frozen Ark Project
has emerged to help set up a more coordinated regime. Its distributed network of frozen gene banks has produced a uniquely cryopolitical mode of
animal management focused on the survival of species and the security
of diversity in subzero suspension. Through the collection of valuable biological material and information and the technical manipulation of temperature, frozen zoos modify the ways species, populations, and organisms
are valued and preserved and create new objects of knowledge and

intervention—epistemic, political, and ontological—in the shape of the cryopreserved genetic species body.

What has become known as "genetic resource banking" involves the freezing of plant and animal genetic material at very low temperatures, using techniques first developed in the mid-twentieth century by, among others, Sir Alan Sterling Parkes, an influential reproductive biologist and a founder of cryobiology. Reproductive material (known as "viable cells") is sampled and stored, both gametes (sperm and oocytes, i.e., eggs) and embryos, and also nonreproductive tissue such as blood and skin, which nonetheless contains an organism's DNA. The goal of such "banking" is to preserve these "genetic resources" for potential future uses: both the reintroduction of valuable genes into dwindling populations at risk of inbreeding or other loss of genetic diversity, and even, technological progress permitting, the possible resurrection of a species after its seemingly inevitable extinction. Such frozen zoos thus serve as supplements to the captive breeding programs of traditional zoos and other institutions (Holt et al. 2003; Watson and Holt 2001; Amato et al. 2009).

The Frozen Ark Project was established by a consortium of universities, research laboratories, conservation groups, zoological gardens, aquaria, and natural history museums to coordinate these cryobanking efforts. It is directed by the University of Nottingham; other participating institutions include the London Natural History Museum's Cryopreservation Unit, the Zoological Society of London, and a number of European, American, and Australian zoos and research centers. With the motto "Saving cells and DNA of endangered species" and the logo of a stable ark on stormy seas, the project's goal is to collect, preserve, and store the genetic material of threatened and endangered animals. Expert panels are tasked with coordinating

Figure 14.1
The Frozen Ark Project's logo.

the various collections and identifying which native and endangered species to sample and the best techniques for each. While it is centered in the Frozen Ark Unit at Nottingham's School of Biology, and is the first and only organized and extensive international program, its goal is not to create a single, secure repository of animal life (a faunal equivalent perhaps of the Svalbard Global Seed Vault, or "doomsday vault"). Rather, it seeks to administer a number of gene banks in different countries, including its own; to share protocols and best-practice techniques for the collection, freezing, labeling, transport, and storage of samples; and to amass and provide member access to a comprehensive online database listing genetic materials stored worldwide.

The Frozen Ark Project should be understood firstly as a counterextinction practice. It acts in a political-ecological context characterized by growing awareness of anthropogenic climate change and widespread, largely destructive, transformation of the natural world—what many are calling the Anthropocene (HARN 2015). Its modes of ethical action are overdetermined by the horizon of the contemporary mass extinction event, that cascading loss of species with multiple, interrelated causes, including habitat loss to development, hunting, war, pastoral overgrazing, introduced species predation or outcompetition, disease, and pollution. Genetic resource banking is one among a suite of counterextinction practices exercising various types of intervention, ranging from *ex situ* breeding, *in situ* management, and habitat protection, to restoration, reintroduction, rewilding, and de-extinction. The Frozen Ark Project is careful to present itself as only one particular venture within this integrated host, as an enterprise that does not replace but contributes to overall conservation efforts by helping maintain genetic diversity in captive breeding and providing a form of *insurance* against genetic loss and even extinction itself. Such conservative pragmatism seeks to rein in the utopian desires often expressed regarding such projects. As the book based on the BBC series *Zoo 2000* breathlessly put it: "It is this, the frozen zoo, that seems to be the great hope for high science, zoos, and animal conservation alike, although it challenges all our notions of what constitutes a zoo" (Cherfas 1984, 98).

Like many other conservationist endeavors, the Frozen Ark Project maintains a rather single-minded focus on endangered species as a moral and practical issue. Beyond a concession to national and community ownership of natural and genetic resources (its sole answer to the question of "ethical issues"), it does little to articulate the cultural and political complexities surrounding extinctions (see Rose, van Dooren, and Chrulew forthcoming).[1] Nor is animal ethics considered relevant; as is common in conservationist discourse, the threat of extinction trumps the rights or welfare claims of individual members of a species. The project seeks rather to extract value (in the form of genetic information) from the material context of both its production and destruction. Its justification is the protection of such valuable genes before they are lost to science and conservation; it seeks also to make possible their reintroduction to help increase genetic diversity. But such salvific suspension is only achieved by separating a secured element of the species from the relational context of animals' lives—their emplacement and duration, their phenomenological worlding, their political, cultural, and ecological milieus. Indeed, a significant part of the value of cryobanking, for some conservationists, is that it can bypass such inconvenient political and material impediments and construct, through the artifice of the archive, a utopian collection of the world's threatened biota: a Frozen Ark in which species, rescued from the catastrophe menacing their places of origin, float free of the encroaching risks of existence.

The Frozen Ark Project is articulated as a form of biopolitical immunization seeking to secure or "freeze" the immemorial dynamism of evolution amid the anthropogenic extinction crisis, to arrest time in hope of a technological or ecological redemption to come. Its website makes the following case for the project's importance:

Despite the best efforts of conservationists, thousands of extinctions have occurred before the animals could be rescued. There has not been enough knowledge or money to stem the tide. This pattern is being repeated across all animal groups and emphasises the importance of collecting the genetic material of endangered animals before they go extinct. The loss of a species allows the results of millions of years of evolution to be lost. If the cells and DNA are preserved, a vast amount of information about a species is saved. Recent scientific developments in molecular biology suggest that in a few decades the recreation of an animal is likely to be possible. The Frozen Ark Project is not a substitute for conservation, but a practical and timely "back-up." (Frozen Ark Project n.d.)

Its discourse combines urgency with fatalism. Faced with bewildering losses and difficult choices in prioritizing resources, with over a thousand species extinctions expected in the next thirty years, and one in every three species classed as threatened, the project's task is apprehended as a race to safeguard something of each species before it disappears forever. Yet this preservable element is not a living community but its DNA, extracted and refrigerated but nonetheless embodied and oriented to the future. As we will see, cryobanking is a practice of *suspension* that freezes genetic information and its vital potentiality in order to secure life against the political and environmental vagaries of living itself.

The Biopolitics of Captive Breeding

The technologies and practices of cryopreservation enact a distinctive twist within the biopolitical regime from which they emerge. The rubric of biopolitics has become an important lens through which to understand how life is subjected to power. In addition to paying prominent attention to the biologization of human politics—in which human beings are apprehended as living animals comprising populations whose health and well-being can be known and managed in ever more exhaustive ways—recent work has explored how animals, too, are subjected to biopower, whether farmed, protected, displayed, rendered, or cloned (Shukin 2009; Chrulew 2012; Friese 2013; Wadiwel 2015). Alongside other productive ordering regimes such as industrial farming, pest and pet control, zoo biology, the "environmentality" of sustainable development (Darier 1999), and wildlife management—and indeed drawing from their expertise—conservation biology has become a dominant apparatus by which the lives of nonhuman animals are administered in the interests of their own survival and salvation.

Zoological gardens have increasingly claimed a significant role within this domain, as an inventive and ever more organized source of techniques for the captive management of wild animals (Chrulew 2010). In the twentieth century, Hediger was preeminently influential in modernizing zoos' operations, placing biological knowledge and intensive nurture at the center of their mission, tailoring forms of care to individual species (Hediger 1964, 1969). Through the development of zoo biology as an applied science combining systematic knowledge with techniques of practical intervention, the zoo directed itself to the task of making wild animals live and, other than protecting them from predators and disease and supplying food, exhibiting the truth of their "natural" behavior in authentic-seeming surroundings. Through collecting and sharing species-specific statistics (from gestation periods to maximum leaps, from optimum temperatures to breeding groups), the conditions of animals' enclosure (such as feeding regime, social structure, exhibit architecture, and enrichment training) could be modified and optimized, the better to satisfy their natural requirements, to foster their lives, and to minimize all artificial disturbances to their behavior and threats to their well-being (Chrulew 2013).

Until the 1960s, conservation was not a high priority in zoological gardens. As public awareness of environmental problems and threats to wildlife increased, zoos increasingly promoted themselves as beacons of endangered species protection, new arks amid eco-catastrophe (Norton et al. 1995; Baratay and Hardouin-Fugier 2002). Increasing legal restrictions on procuring animals from the wild or via trade, collection practices on which zoos had previously relied, and the enshrining of conservation goals in the regulations of their accrediting bodies, presented new problems and opportunities for their intensive methods of husbandry, and led to the development of breeding programs aimed at self-sustaining zoo populations (Donahue and Trump 2006). As it had within human biomedicine and animal agriculture, sexual reproduction—the meeting point of the body and the population, the respective targets of disciplinary power and biopower—became a central object of scientific knowledge and intervention in captive animals' lives (Clarke 1998; Twine 2010). Where once serendipity had largely reigned, there developed a highly organized, rationalized system of population management through studbooks, taxon advisory groups, and other forms of regional and international cooperation—a nonhuman *scientia sexualis* sharing animals and their body parts as well as birth and death records, demographic and fertility statistics, gamete collection and insemination techniques, and other procedures for the breeding of exotic species. The 1970s and '80s saw a great increase in successful captive breeding. Since the American Zoo and Aquarium Association declared as a major priority "becoming the leader in the preservation of rare and endangered animals," Species Survival Plans (whose original title was "master breeding plans") have guided the exchanges of animals according to criteria of genetic diversity and sustainable demography, seeking eventually to support wild populations through reintroduction (Chrulew 2011; American Zoo and Aquarium Association n.d.).[2]

Insofar as it sought to respond not merely to the deficiencies and dangers internal to an otherwise self-sustaining population, but to the possible *extinction* of that very population itself, conservation biology performed a distinctive intensification of zoological biopower. As Foucault argued, bio-politics addresses itself to mortality, to the preventable yet "normal" disor-ders and aleatory events that threaten the lives of individuals and groups: "At the end of the eighteenth century, it was not epidemics that were the issue, but something else—what might broadly be called endemics, or in other words, the form, nature, extension, duration, and intensity of the ill-nesses prevalent in a population" (Foucault 2003, 243). In the twentieth century, of concern were not only such endemic disorders, but also *extinc-tions*, events that threatened life itself, the wholesale loss of a population or species. Humankind became seen as itself an epidemic of sorts, tasked with preventing its own worst impacts on the natural world. Yet just as death was beyond power's reach, so too was extinction: it was still only mortality, and the details of threatened animals' lives, that could be known and inter-vened upon. Biopolitical counterextinction practices here addressed them-selves not only to the internal health of a population but to its very grounds of reproduction and perpetuation, its capacity for generation.

The Epistemology of the Postvital Life Sciences

Genetic resource banking was made possible by the genetic revolution in the twentieth-century life sciences. The discovery of DNA, and the tech-niques for sequencing and manipulating it, transformed the scope of biol-ogy epistemologically, politically, and indeed ontologically. If the opening up of corpses following the development of autopsy made visible a new realm of disease and death (Foucault 1994), and if the development of biol-ogy as a discipline looked beyond the surfaces of bodies to the vital func-tions of organs (Foucault 2002), then twentieth-century genetics dug deeper into the patterns that make up these newly penetrated bodies, its biomedia seeing beyond the organic space of the body to the informatic space of its constitutive chromosomes. The revelation of life as information, as lan-guage, produced a new form of knowledge, attentive to novel entities beyond the anatomy and behavior of the organism. This distinctive gaze generated new technologies of intervention into the fabric of life, at the same time as it transformed techniques of power over animal bodies and populations (Gigliotti 2009).

Richard Doyle's *On Beyond Living* examines the epistemological break from the organism to the gene marked by the rise of molecular biology and other twentieth-century life sciences. As Foucault showed, the historical transformation from natural history to biology involved an epistemic shift from the taxonomy of characteristics that are visible and comparable, to the study of "life" as an essential yet invisible element of the organism (Foucault 2002). Doyle recounts biology's shifting attention beyond visible surfaces to the functional object of "organic structure," and the fundamental ground of classification, "life," as that concealed within the organism, "a concept beyond the particularities and practices of living organisms" (Doyle 1997, 12). He also traces the subsequent shift from the vitalist object of "life" to the object of "the 'postvital' body ... [which] fits, and is fitted to, molecular biology" (8). This is "a body in which the distinct, modern categories of surface and depth, being and living, implode into the new density of coding," the information of DNA (13). He argues that Schrödinger's rhetorical shift from organism to code, phenotype to genotype, is what made molecular biology possible: "With this move—the metonymic substitution of 'code' for 'organism'—the entire future birth, life, and death of the organism is 'contained' or engulfed by the chromosomes" (28). This new scientific way of seeing looks beyond the organism to fixate on internal codes.

Lily Kay has also traced the discontinuous history of the genetic code trope of molecular biology, its emergence alongside information and computer sciences and cybernetics, extending across three interactive periods—the "specific," "formalistic," and "biochemical" phases—into a "language" proper that needed to be cryptographically cracked (Kay 2000; Kay 1993; Keller 1995). To these processes of "metaphorization," Eugene Thacker adds the "biotechnical" and "bioinformatic" phases, whose equation of DNA with information saw the rise of genetic engineering and the integration of biotechnology with computing in the era of genome databasing (Thacker 2004, 37; 2005, 2010). In these latter phases, which provide the epistemic and technical context for cryobanking and frozen zoos, "information is seen as constitutive of the very development of our understanding of 'life' at the molecular level—not the external appropriation of a metaphor, but the epistemological internalization and the technical autonomization of information as constitutive of DNA" (Thacker 2004, 40). In the rhetorical modes of the contemporary life sciences, biology's object of "life" is transformed: organisms and their hidden life give way to DNA beyond life, and corporeal living beings are often seen less as subjects of worlds in evolving community than as expressions of code, temporary vessels of information.

This new scientific discourse also changed the significance of extinction. The loss of the last remaining individual had routinely been the locus through which the extinction of a species was articulated, and intense efforts were focused on capturing or rescuing (in photographic or scientific media) the final surviving population, to prevent the irredeemable loss of a unique quality, a singular mode of being in the world. With molecular biology, the species concept—the singular form expressed in each individual organism—became rearticulated in terms of DNA, which took on a "status as an immanent, organic, eternal form" (Doyle 1997, 81). Extinction was thus conceivable as a loss not only of a population but of information. Yet as information can be discovered and stored in sites other than bodies, or in bodies in states other than living, a species might thus be located beyond even the last survivor of its kind. This relocation of life onto DNA that can be stored and manipulated in computational or cryogenic archives made the previously impenetrable void of extinction a knowable and potentially redeemable domain. As Stephanie Turner argues, in the conceptual framework of genetic science the black hole of extinction becomes "open-ended"; what was once articulated as a long, irreversible evolutionary process is now seen as an endless stream of data, of permutations subject to human control: "In genome time, evolutionary histories, including extinction narratives, are revised, forestalling or even reversing absolute endpoints in the endless reproducibility of the DNA code" (Turner 2007, 59). Yet as well as making possible the remarkable de-extinction projects she discusses, "genome time" also changes the parameters of more conventional counterextinction practices, insofar as storing genetic information is seen to constitute an important part—perhaps in some ways even a sufficient one—of "saving" a species.

The Temporality of Cryopreservation

Do frozen zoos then simply exemplify reductive technoscience, insofar as they isolate and preserve biological materials containing genetic information rather than living organisms in community and generational continuity? Are these banks of genetic resources merely the latest enframing technology, reducing the world of animal being and becoming to "standing reserve" (Heidegger 1977)? Certainly, they significantly transform their objects, addressing the world as a realm of archivable information and material, and deploying extreme temperatures to suspend a savable remnant. As Parkes observed, once viable sperm had first been successfully recovered after freezing, "time and space had been abolished" (quoted in Holt et al. 2003, 269). Yet it is important to address the complexity of the remediations that frozen zoos perform. Their encoding and decoding of bodies, their sampling and preservation, manipulate temporality, embodiment, and emplacement in novel ways, and produce new scientific and political objects, new entities and relationships.

The Frozen Ark Project indeed generates unprecedented objects of knowledge and intervention in and through its cryogenic archives. The practice of captive breeding had already fabricated a new type of population: rather than a group of cohabiting animals who share a relatively coherent time and place, the transport of animals between zoological gardens and the translocation of wild-living groups enabled effective breeding populations to exist despite extreme spatial dislocation, resulting in uncannily dispersed yet artificially integrated reproductive communities. Biotechnologies further reshaped captive breeding logistics; in the zoo as in the farm and the kennel, the ability to ship semen rather than animals separated the process of insemination from the need for copulation, with transformative economic, ethical, and ontological results.[3] The postvital life sciences and their technologies thus rendered the rather nebulous objects of "species"—the subject of extensive philosophical and scientific debate over whether they exist and if can we have obligations to them—newly visible, orderable, and amenable to modification and manipulation. The archiving of genetic material served to *double* captive populations that already doubled those in the wild, further displacing their preservation to the freezers of the lab. Through the distributed collection of frozen tissues, cryobanking produces and acts upon a virtual species body in new ways.

Yet this archive, while thoroughly technologically mediated, remains uncannily biological. It is not static but anticipatory; while located *beyond living*, it is nonetheless *lively*, devoted to life, composed in relationships with living populations, the apparatus that maintains it, and the future it seeks to enliven. Life is never entirely reducible. As Doyle argues in his analysis of the parallel practice of human cryonics: "If molecular biology fostered a forgetting of the body and a remembering of code, DNA, then cryonics enables a return of the body, the body as code" (Doyle 2003, 75; for more on cryonics, see Parry 2004; Bunning, this vol.). Frozen zoos are not archives of biological data in a computational sense, nor solely archives of material containing DNA, but also collections of reproductive material, of gametes and embryos saved for future thawings, inseminations, and reintroductions into living populations, intended to invigorate their diversity and resilience. Like the cryonic human bodies that, in becoming frozen, seek to transcend death, the cryogenic species bodies are constituted by a promise to preserve the ark itself, the bodies that it contains, and (ideally) the wild populations they represent: "At some date in the future, when the captive population is too inbred, the embryo bank would be tapped to resurrect some of the old genes, thus restoring its variability and vitality" (Cherfas 1984, 100). The frozen zoo seeks not only to hide away, to store or secure, but also to *make possible*, to enable the very future of life— although not necessarily in the same form as that from which it was extracted. It seeks to preserve potentiality, diversity, variability, and adaptability, in a reproducible, reconstitutable way: "The stored cells and gametes provide a renewable resource of variation for revitalising breeding programmes" (Frozen Ark Project n.d.). It freezes cells in their most plastic and pluripotent form, securing the potentiality of latent life and anticipating and maximizing as yet unknown future uses (Landecker 2007; Radin 2015).[4] It enrolls scientists and animals in a network of responsibility, tied not only to existing others, to presently living individuals, but to species as well, and bound proleptically to the future generations of those species or what they will become. As Doyle puts it, "The frozen body does not stand still" (2003, 64): it is both living and latent, suspended in and out of time, both present now and existing in a promissory relationship with the future, seeking to secure the conditions of possibility of the (nonhuman) other to come.

Yet while it is itself unescapably at risk, conditioned by and exposed to the future, the frozen zoo nonetheless seeks to insure this very future and secure it against risk. Most of the future conditions that the Frozen Ark Project anticipates—the renewal of habitat and survival of threatened populations, after the floodwaters of anthropogenic eco-apocalypse subside— are discounted as beyond its control; instead it devotes itself intensely to a particular set of technical undertakings dedicated to maintaining its specimens in perpetuity. Like the individual human subject of cryonics, the animal species "has entered into a relation with a future ... a massive realm of incalculable contingency that must be continually managed, even disciplined" (Doyle 2003, 66). The freezing of the genome goes hand in hand with the intensive management of its target population, to ensure that they are ready for and adequate to the future gift that cryobanking has promised them. Moreover, in seeking to preserve species against threats to their survival, it alters the very nature of their perpetuation. Like the zoo itself, only more so, it transforms living beings' temporality, their relationships to ancestral labor, their mode of generational transmission.

The Cryopolitics of Frozen Zoos

At the same time as cryopower mediates the virtual species body beyond living, it multiplies very material interventions into animals' lives. Genetic resource banking further expanded the reach of conservation biology and captive breeding into new hi-tech domains. While, as Doyle puts it, the gaze of these postvital life sciences "does not see bodies; it sees only sequences, genomes," it nonetheless *"requires"* bodies for its operation (Doyle 1997, 131, emphasis added). As the means by which to manipulate the object of knowledge that is the genetic species body, the "invisible" living bodies of animals remain the ever more starkly visible objects of biopolitical shepherding.

Over twenty years ago, in a book lauding zoos as conservation leaders, Colin Tudge enthusiastically analyzed their use of biotechnology and the rise of frozen zoos (Tudge 1992, 169–192). He expressed his excitement that this was possible and his frustration that so few places were doing it, alongside a vague unease at potential moral dilemmas, easily quashed by the overwhelming justification of saving species from extinction. As Tudge made clear, *"There is an infinity of possible interventions,* to correct malfunction at any point along the way, or to enhance processes that are already healthy and natural, but do not serve the present needs of other mammalian populations that are dwindling rapidly towards extinction" (171, emphasis added). That is, sexual reproduction is an opportunity for radically multiplying interventions to enhance natural and normal processes for conservation ends. The expanding range of technologies then included boosting fertility, artificial insemination, egg rescue, in vitro fertilization, embryo transfer and storage, chimaeras, and cloning.[5] These require interventions of varying intensity, described with typical scientific neutrality, from anaesthetization to the manual collection of semen. The technical difficulty yet ultimate convenience of predicting and controlling ovulation cycles means that rigorous monitoring is required. Hormonal changes are measured in blood, urine, and feces, and through ultrasound. Separation of offspring and subsequent hand-rearing is common, often leading to "behaviorally incompetent" individuals incapable of surviving in the wild or rearing their own young, further perpetuating the cascade of intervention (see further van Dooren, this vol.; Chrulew forthcoming). The Frozen Ark Project encourages opportunistic sampling of species during veterinary treatment and wildlife tagging, hoping it will become a normalized expectation of zoos as well as in *in situ* conservation programs. As with new medical technologies, from fetal screening for prenatal diagnosis to genetic testing for inherited vulnerabilities, this new domain of knowledge becomes not only problematized as a domain of possible intervention but indeed moralized as a domain of necessary intervention. Given rates of extinction, it would be "criminal" not to store the genes of all animals. Ultimately the development of cryopreservation means the further multiplication of demands for intervening in the lives of wild and captive animals according to genetic norms.[6]

At the same time as they assist captive breeding, frozen zoos also transform its mode of operation, in particular by making possible the storage and manipulation of a species body divorced from any living members of the population. Their unique compression of space and suspension of time here diverges from the prevailing zoo-biopolitical logics that they also supplement and intensify. They collect and order not a living spectacle for public exhibition but a set of functional biological artifacts. A frozen zoo has no visitors or displays; it can exclude the already embarrassing justifications of entertainment and education and concentrate entirely on research and conservation (Lee 2005). Alongside the conservationist care that targets the reproduction and flourishing of a surviving population, whether *in* or *ex situ* (or increasingly, via metapopulation management, both), frozen zoos focus on materials extracted from and, for certain uses, substitutable for living organisms. They thus eliminate the necessity of taking animals from the wild to boost the captive population's genetic diversity, or of transferring animals between zoos for planned matings, a risky practice due not only to the difficulties of transport—a stressful and dangerous procedure that regularly leads to injury, trauma, or fatality—but also to the inscrutability of animal agency in sexual encounters (Hediger 1969, 217–243). The genes of dead animals can even be reintroduced into the population, circumventing the limits of age and death and reorganizing patterns of generational transmission. The transfer of cryopreserved cells thus allows zookeepers to collect and exchange genes and to manage reproduction while bypassing the risks to conservation and welfare of collecting or transporting animals, the frustrating vagaries of mate incompatibility, and the loss of genetic diversity through mortality. Yet this development of noninvasive or simply more efficient interventions, which has its own economic and ethical advantages, at the same time reveals the inadequacy of zoo care for producing "natural" reproductive behaviors (Hediger 1986).

Indeed one outcome of the frozen zoo is to minimize the necessity of managing living animals at all. Once sampled, an individual animal is no longer required for *its own* reproductive success. Effective breeding

by natural reproduction involves maintaining large numbers of breeding individuals usually dispersed amongst geographically disparate zoos. Storing germplasm samples from many individuals releases some of the strictures on large space requirements. Fewer living animals are necessary, and more space becomes available for other rare species. (Holt et al. 2003, 267–268)

Not only inferior specimens but all individuals become potential surplus to the reproductive needs of their species. Once sampled, the living become alienated from their own genetic resources, and the labor of managing and caring for their lives becomes unnecessary and dispensable.

The new forms of preservation and exchange made possible by the frozen zoo transform the relationships between humans, animals, and technologies, reorganizing space and time beyond familiar constraints in the interests of optimal efficiency and diversity. Animals no longer need be present or even alive to breed. *Fewer living animals are necessary*. As the benchmark zoo management textbook puts it: cryobanking "represents a technically viable method for helping to conserve species biodiversity, without having to maintain large captive populations of each organism" (Hosey, Melfi, and Pankhurst 2009, 319; see also Ryder and Benirschke 1997; Friese 2013, 52). In this sense, cryopolitics significantly transforms and intensifies the methods, priorities and obligations of the biopolitics from which it emerges. Zoos of course already perform their own transposition of habitats and territories, designing numerous etho- and zoo-technologies that substitute for nature in reproducing animals' wildness. Frozen zoos amplify, even culminate this displacement. The accumulated knowledge and techniques devoted to making zoo animals live—taking into account their species specific requirements in enclosure, feeding, breeding, enrichment, and care—now operate upon smaller and more optimally rationalized populations composed of both living and nonliving objects. As Friese puts it, "By linking frozen zoos with SSPs [Species Survival Plans], endangered species are increasingly being thought of as including both embodied individuals (where genetic information is materializing) and cryopreserved cells (where genetic information is suspended in a liminal state)" (2013, 109–110). The use of extreme temperature to freeze reproductive materials enables the isolation of genetic potentiality from emplaced and embodied actuality, the separation of the species body from the living bodies of individuals and groups. The administration of this bioarchive involves the protocol of the lab rather than the pragmatic husbandry of the zoo, let alone the management of the "wild." It replaces the subjectification of dependents with modes of technical objectification. The cryopolitical domain, that which frozen zoos "have to maintain," is no longer the milieus of living beings, but the liminal space of the freezer.

Indeed, in suspending the temporal relationship of a species to its vital context, cryopreservation exacerbates the disruption already enacted in zoo biology practices that separate animals from their social and natural environments and surround them with a wholly anthropogenic apparatus of care. In their technological utopia, frozen zoos rescue life by bypassing the hazardous, aleatory, and agonistic dimensions of living itself. By freezing time, arresting the corruption of the species' genetic code, all the vagaries of both nature and history are held at bay. The risks associated with breeding over generations such as genetic loss and drift can be circumvented by freezing select resources and reintroducing their diversity, unsullied by time, into future generations. The dangers associated with managing the security of a wild population—political threats, habitat loss, disease, poaching—can likewise be averted. Here living itself, exposure to time and the other, becomes a risk to be avoided. Yet just as generation and corruption, coming into being and passing away, are central to the concept of life in Western thought, "the idea of extinction doubly haunts every instance of the living, as an individualized living being and as a generalized type or form of life" (Thacker 2010, 25). All life, all individuals and all species, are subject to decay, and to the risk of extinction. At its extremes, biopower strives to secure life against not only specific instances of disease and decay but against the very possibility of degeneration, risk, and loss, against time and death as such. Faced with the qualitatively amplified threat of extinction, of species death, cryopower strives to secure life against *living itself.*

In doing so, cryopower transfigures its object, maintaining it in suspended animation, as life's code or potential but no longer as living, without an experiential relationship to the world, or an autonomous capacity for communal generation. It seeks to save life, here not through all the effort of shepherding the living through dangers and generations in their everyday needs, but through largely technical means. If, as Foucault put it, sovereign power *makes die* and *lets live*, and biopower *lets die* and *makes live*, then cryopower refuses even to *let die*—at least at the level of the gene and species. Instead it intervenes at the extremes of life in order to prevent death *qua* extinction. But—and this is the irony of cryopolitics—in doing so it seeks to protect life from much that essentially characterizes living: its embodiment, its relationships to the world, to the environment, to ancestors and offspring, to other creatures, to human beings, to the passage of time, to reproduction and intergenerational transmission. Rebuilding such relationships remains possible—indeed, it remains conservation's very task. Amid multiple migrations and transformations of techniques and biological materials—between zoo and wild as much as between zoo and lab—it remains to be seen what effect the frozen ark might have on the other vessels in the zoological flotilla. But there is an important sense in which cryopower risks becoming an excessive immunization of life, a "negative [form] of the protection of life" that "saves, insures, and preserves the organism, either individual or collective, to which it pertains, but it does not do so directly, immediately, or frontally; on the contrary, it subjects the organism to a condition that simultaneously negates or reduces its power to expand" (Esposito 2008, 46). In seeking to protect life from the vagaries of living itself, cryopreservation takes biopower to its postvital extreme.

Questionable Futurology

The eminent zoo director Hediger regularly spoke out against the encroachment of industrial and laboratory practices and technologies within the zoo. In his 1986 article "The Zoo in the Fridge: Questionable Futurology," he looked back over the development of zoo biology in the last half-century, at the stages of animal husbandry that his career had overseen, such as when a species was first imported or successfully bred. What is the virtue of the frozen zoo, he asked, as a contribution to zoo biology, that is, to the maintenance of natural animals? What new stage of animal husbandry does it engender? He saw cryobanking as largely another instance of technical advancement at the expense of proper care: "Both domestication and laboratorization must be kept away from the zoo, as I have always emphasized."[7] He resisted such objectification in favor of his own sophisticated ethopolitical techniques for conducting animal behavior: "Of course, all the aforementioned technical breeding tricks are, in a sense, significant progress; but they are technical progress—focused on biological objects, which should be rather treated as subjects, namely, as respectable, sentient individuals, as precious bearers of life" (88). In particular, cryopreservation can only conserve fixed genetic behavior and not that transmitted through cultural learning: "Everything based on tradition is lost in pure refrigerator breeding, and this is perhaps not a little" (86).[8] For Hediger, frozen zoos could only be an annex to the proper maintenance of living populations; indeed, "It is also a dead end, because its contents and products—and be it ever so precious seeds, eggs, embryos, etc.—are worthless without the supplier and the recipient maintained under near-natural conditions, as we find them only in the traditional zoo and the dwindling remnants of so-called wildlife" (90).

Perhaps the displacements and objectifications of the frozen zoo reveal, a little too uncomfortably, those already present in zoos themselves. And no doubt the lab is more lively than this zoo man saw. He barely glimpsed the extent to which, as part of a suite of counterextinction practices, zoos would today become intimately and productively entwined with the laboratories and wildlife habitats they had long defined themselves against. Perhaps too cryopreservation enables less the takeover of the zoo by the lab that Hediger feared than the zoo's appropriation of the lab for its own distinctive purposes. It is in any case no exaggeration to say that the future conditions of existence of numerous forms of life rest on the outcomes of these troubled and creative negotiations between zoo, lab, and wild. How will living animals persist, both cared for in the optimized temperatures of their artificial enclosures, and exposed and adapting to the changing climates and various dangers of their habitats? In stainless steel cylinders, their frozen remnants await future uses.

Notes

1. "What about ethical issues? We respect the rights of local communities and national governments over their natural resources, including genetic resources, and will protect them. Rules governing the export of samples from their countries of origin will be observed, as will procedures to prevent harm when taking samples" (Frozen Ark Project n.d.).

2. On the politics of the creation of Species Survival Plans by zoo associations, see Donahue and Trump 2006, 108–139. The effectiveness of zoos' captive breeding programs for endangered species preservation has often been questioned. Zoos lack the resources to hold many threatened species, and when they do, the populations are often not adequately self-sustaining.

3. Indeed the development of cryopreservation in zoos drew on expertise and technology from industrial agriculture, although, as many have pointed out, while the animal industries are motivated by profit to inbreed for desirable traits, zoological cryobanking rather resists domestication, outbreeding to maintain the biodiversity of viable populations.

4. See Radin 2015 on "planned hindsight," the "strategy for choreographing life, time, and value" (361) that is enacted in the cryopreservation of biospecimens.

5. As well as interspecies embryo transfers, there have been a number of cases of transspecies cloning, the first being the birth to the domesticated gestational surrogate of "a humble Iowa cow" of a gaur named Noah who survived for only two days. Lanza, Dresser, and Damiani (2000) hoped that "Noah will be just the first creature up the ramp of the ark of endangered species." For more on controversies regarding the cloning of endangered species in zoos, and the way in which its various models articulate different imaginaries of wild life and make nature in different ways, see Friese 2013.

6. The moderation of disturbance is here an open question, as increased intervention undermines the goal of naturalness. Zookeepers will generally prefer not to intervene unnecessarily. Yet biologically sound management demands precise and extensive knowledge, gained only through forms of more or less invasive contact with the animals. The history of zookeeping thus consists in ingenious attempted solutions to these dual demands: forms of noninvasive surveillance and self-erasing intervention that at times resemble merely the latest attempt at disembodied objectivity, but often also demonstrate a perceptive awareness of their unique situatedness—techniques of engaged invisibility and arts of minimal disruption.

7. Many thanks to Christian Kroos for his help with translations from this article.

8. Of course, zoos have their own problems with achieving their goal of retaining natural behavior, often overestimating the effectiveness of their artificial environments and enrichments (see, e.g., Chrulew forthcoming). But the cryopolitical focus on biological objects only further overvalues genetic inheritance in relation to epigenetics, gestation, nurture, subjectification, and the cultural transmission of learned behaviors. Friese, for example, explores the concern that "the frozen zoo does not provide a way to regenerate the behaviors of those animals, which to many people's mind is crucial in delineating what a species is and how it will survive into the future" (2013, 176). How such genetic reductionism relates to other forms of behavioral objectification that persist in zoos demands further analysis.

References

Amato, George, Rob Desalle, Oliver Ryder, and Howard C. Rosenbaum, eds. 2009. *Conservation Genetics in the Age of Genomics*. New York: Columbia University Press.

American Zoo and Aquarium Association. n.d. http://www.aza.org/ (accessed September 22, 2014).

Baratay, Eric, and Elisabeth Hardouin-Fugier. 2002. *Zoo: A History of Zoological Gardens in the West*. Trans. O. Welsh. London: Reaktion Books.

Bowkett, Andrew E. 2009. Recent captive-breeding proposals and the return of the ark concept to global species conservation. *Conservation Biology* 23 (3): 773–776.

Cherfas, Jeremy. 1984. *Zoo 2000: A Look Beyond the Bars*. London: British Broadcasting Corporation.

Chrulew, Matthew. 2010. From zoo to zoöpolis: Effectively enacting Eden. In *Metamorphoses of the Zoo: Animal Encounter after Noah*, ed. Ralph R. Acampora, 193–219. Lanham: Lexington Books.

Chrulew, Matthew. 2011. Managing love and death at the zoo: The biopolitics of endangered species preservation. *Australian Humanities Review* 50:137–157.

Chrulew, Matthew. 2012. Animals in biopolitical theory: Between Agamben and Negri. *New Formations* 76:53–67.

Chrulew, Matthew. 2013. Preventing and giving death at the zoo: Heini Hediger's "Death Due to Behaviour." In *Animal Death*, ed. Fiona Probyn-Rapsey and Jay Johnston, 221–238. Sydney: Sydney University Press.

Chrulew, Matthew. Forthcoming. Saving the golden lion tamarin. In *Extinction Studies: Stories of Time, Death, and Generations*, ed. Deborah Bird Rose, Thom van Dooren, and Matthew Chrulew. New York: Columbia University Press.

Clarke, Adele. 1998. *Disciplining Reproduction: Modernity, American Life Sciences, and "the Problems of Sex."* Berkeley: University of California Press.

Corley-Smith, Graham E., and Bruce P. Brandhorst. 1999. Preservation of endangered species and populations: A role for genome banking, somatic cell cloning, and androgenesis? *Molecular Reproduction and Development* 53:363–367.

Darier, Éric, ed. 1999. *Discourses of the Environment*. Oxford: Blackwell.

Donahue, Jesse, and Erik Trump. 2006. *The Politics of Zoos: Exotic Animals and Their Protectors*. DeKalb: Northern Illinois University Press.

Doyle, Richard. 1997. *On Beyond Living: Rhetorical Transformations of the Life Sciences*. Stanford: Stanford University Press.

Doyle, Richard. 2003. Disciplined by the future: The promising bodies of cryonics. In *Wetwares: Experiments in Postvital Living*, 63–87. Minneapolis: University of Minnesota Press.

Esposito, Roberto. 2008. *Bíos: Biopolitics and Philosophy*. Trans. T. Campbell. Minneapolis: University of Minnesota Press.

Foucault, Michel. (1963) 1994. *The Birth of the Clinic: An Archaeology of Medical Perception*. Trans. A. M. Sheridan Smith. New York: Vintage Books.

Foucault, Michel. (1966) 2002. *The Order of Things: An Archaeology of the Human Sciences*. London: Routledge.

Foucault, Michel. (1997) 2003. *"Society Must Be Defended": Lectures at the Collège de France, 1975–1976*. Ed. Mauro Bertani and Alessandro Fontana. Trans. David Macey. New York: Picador.

Friese, Carrie. 2013. *Cloning Wild Life: Zoos, Captivity, and the Future of Endangered Animals*. New York: NYU Press.

Frozen Ark Project. n.d. http://www.frozenark.org (accessed September 22, 2014).

Gigliotti, Carol, ed. 2009. *Leonardo's Choice: Genetic Technologies and Animals*. Dordrecht: Springer.

HARN Editorial Collective, eds. 2015. *Animals in the Anthropocene: Critical Perspectives on Non-Human Futures*. Sydney: Sydney University Press.

Hediger, Heini. (1942) 1964. *Wild Animals in Captivity: An Outline of the Biology of Zoological Gardens*. Trans. G. Sircom. New York: Dover.

Hediger, Heini. (1963) 1969. *Man and Animal in the Zoo: Zoo Biology*. Trans. G. Vevers and W. Reade. New York: Delacorte Press.

Hediger, Heini. 1986. Der Zoo im Kühlschrank: Fragwürdige Futurologie. *Der Zoologische Garten (N.F.)* 56 (2): 81–90.

Heidegger, Martin. 1977. *The Question Concerning Technology and Other Essays*. Trans. W. Lovitt. New York: Garland.

Holt, William V., Teresa Abaigar, P. F. Watson, and David E. Wildt. 2003. Genetic resource banks for species conservation. In *Reproductive Science and Integrated Conservation*, ed. William V. Holt, Amanda R. Pickard, John C. Rodger, and David E. Wildt, 267–280. Cambridge: Cambridge University Press.

Hosey, Geoff, Vicky Melfi, and Sheila Pankhurst. 2009. *Zoo Animals: Behaviour, Management, and Welfare*. Oxford: Oxford University Press.

Kay, Lily E. 1993. *The Molecular Vision of Life: Caltech, the Rockefeller Foundation, and the Rise of the New Biology*. New York: Oxford University Press.

Kay, Lily E. 2000. *Who Wrote the Book of Life?* Stanford: Stanford University Press.

Keller, Evelyn Fox. 1995. *Refiguring Life: Metaphors of Twentieth-Century Biology*. New York: Columbia University Press.

Landecker, Hannah. 2007. *Culturing Life: How Cells Became Technologies*. Cambridge, MA: Harvard University Press.

Lanza, Robert P., Betsy L. Dresser, and Philip Damiani. 2000. Cloning Noah's Ark. *Scientific American* (November): 84–89.

Lee, Keekok. 2005. *Zoos: A Philosophical Tour*. Basingstoke: Palgrave Macmillan.

Norton, Bryan G., Michael Hutchins, Elizabeth F. Stevens, and Terry L. Maple, eds. 1995. *Ethics on the Ark: Zoos, Animal Welfare, and Wildlife Conservation*. Washington, DC: Smithsonian Institution.

Parry, Bronwyn. 2004. Technologies of immortality: The brain on ice. *Studies in History and Philosophy of Biological and Biomedical Sciences* 35 (2): 391–413.

Radin, Joanna. 2015. Planned hindsight. *Journal of Cultural Economy* 8 (3): 361–378.

Rose, Deborah Bird, Thom van Dooren, and Matthew Chrulew, eds. Forthcoming. *Extinction Studies: Stories of Time, Death, and Generations*. New York: Columbia University Press.

Rose, Nikolas. 2007. *The Politics of Life Itself: Biomedicine, Power, and Subjectivity in the Twenty-First Century*. Princeton, NJ: Princeton University Press.

Ryder, Oliver A., and Kurt Benirschke. 1997. The potential use of "cloning" in the conservation effort. *Zoo Biology* 16 (4): 295–300.

Shukin, Nicole. 2009. *Animal Capital: Rendering Life in Biopolitical Times*. Minneapolis: University of Minnesota Press.

Soulé, Michael E. 1991. Conservation: Tactics for a constant crisis. *Science* 253:744–750.

Thacker, Eugene. 2004. *Biomedia*. Minneapolis: University of Minnesota Press.

Thacker, Eugene. 2005. *The Global Genome: Biotechnology, Politics, and Culture.* Cambridge, MA: MIT Press.

Thacker, Eugene. 2010. *After Life*. Chicago: University of Chicago Press.

Tudge, Colin. 1992. *Last Animals at the Zoo: How Mass Extinction Can Be Stopped.* Oxford: Oxford University Press.

Turner, Stephanie S. 2007. Open-ended stories: Extinction narratives in genome time. *Literature and Medicine* 26 (1): 55–82.

Twine, Richard. 2010. *Animals as Biotechnology: Ethics, Sustainability, and Critical Animal Studies*. London: Earthscan.

Wadiwel, Dinesh Joseph. 2015. *The War Against Animals*. Leiden: Brill.

Watson, P. F., and W. V. Holt, eds. 2001. *Cryobanking the Genetic Resource: Wildlife Conservation for the Future?* London: Taylor & Francis.

15   The Utopia for the Golden Frog of Panama

Eben Kirksey

Utopian worlds are ever-present in science fiction, where the projection of new heavens is never far from the emergence of new hells (Williams 1978, 212). Utopias can also function as diagnostic tools. To paraphrase Isabelle Stengers, they are learning grounds for resisting what today opportunistically frames our world (Stengers 2011, 347). In collaboration with Grayson Earle, a digital artist, and Mike Khadavi, a frog enthusiast who designs custom aquariums, I attempted to create a learning ground—an artwork called *The Utopia for the Golden Frog of Panama*—to diagnose cryopolitical problems associated with biodiversity conservation initiatives. Around the world, species of frogs are dying. A global assemblage of biosecure holding facilities and cryogenic banks called the Amphibian Ark has been built to save species of frogs that cannot currently exist in the wild.[1] Thousands of frogs will remain within the Ark, a place where science fictions meet Christian messianic traditions, until circumstances change (cf. Haraway 2014). This chapter uses the fate of one particular species, the golden frog of Panama, to examine the politics of efforts to forestall extinction.

At the register of materials, *The Utopia for the Golden Frog of Panama* was a repurposed refrigerator with a window enabling viewers to observe endangered frogs as they were subjected to the ambivalent grace of salvation. Our "utopia" was pregnant with irony. Living within a refrigerator, in conditions of incarceration, is certainly not utopic, not even for a frog. Our goal in creating this art installation was to catalyze conversation about the genuine ethical and logistical difficulties that emerge when one grapples with endangered life forms in the early twenty-first century. Taking a page from *Tactical Biopolitics*, a 2008 book by Beatriz da Costa and Kavita Philip that brings the ideas of Michel Foucault into conversation with tactical media practices in the arts, we made a concrete proposal for managing life and death differently (da Costa and Philip 2008). Rather than just offer a critique of standard zoological practices, this installation served as an opening to possible futures.

The Utopia for the Golden Frog of Panama also offers a figural window into the Amphibian Ark. Our intervention aimed to expose strategies for managing life and death within zoological facilities (see da Costa and Philip 2008; Catts and Zurr 2008, 131). By placing ethnographic descriptions of the Amphibian Ark alongside an account of the creation of our own humble utopia, this essay chronicles the actions of people whose love for some kinds of life has led them to construct novel ecosystems—bringing machines, industrial supply chains, and biological elements together into unusual assemblages (see also Kirksey 2015). Within Amphibian Ark facilities I found people committed to the practical work of care whose imaginations were constantly probing future horizons (see van Dooren 2014b; Crapanzano 2004). As the Ark ran out of space, as zoo keepers started killing frogs to keep populations manageable, caretakers imagined possible futures where the animals might escape their present circumstances.

The Year of the Frog (Panama, December 2008)

A comfortable breeze from the air conditioner, a steady 24 degrees Celsius, hit the zoo keeper in the face as she was greeted by the familiar burbling from the aeration tubes and hum of the air pumps. The slash/chink rhythm of a machete, cutting the grass outside, was still audible over the automated systems of the El Valle Amphibian Conservation Center (EVACC, pronounced like *evac-uation*). She washed her hands with antibacterial soap in the sink and put on a pair of powderless rubber gloves. At the top of a new page in the log, a cheap spiral notebook, she wrote: "21 December 2008, *Atelopus* room." Peering into the first couple of tanks, she found little to report in the log: "0 fecal, 0 food left, one frog on plant, one hiding in peat moss." Spritzing each tank with a blast from the garden hose, with the nozzle turned to the *mist* setting, she quickly moved down the row.

EVACC is "a space age amphibian center nestled in the heart of an extinct volcanic crater" in the highlands of Panama, in the words of Lucy Cook, who writes the Amphibian Avenger blog. She describes EVACC as "a terrifying vision of the future where frogs survive in sterile pods and crocs

Figure 15.1
Heidi Ross amid her daily routine of care at EVACC in El Valle, Panama. Photograph
by Lucy Cook.

are mandatory footwear" (Cook 2010).[2] The Amphibian Avenger blog is
dedicated to "the ugly, the freakish and the unloved animals that are peril-
ously ignored thanks to the tyranny of cute" (ibid.). When Lucy Cook visited the EVACC facility, she was delighted to find "freakish" species living
alongside cute frogs.

During my own visit to EVACC, I found human caretakers who had
become emotionally and ethically entangled with creatures in their care,
people who were committed to the practical labor of keeping frogs alive
and helping them flourish in an era of extinction. Rather than terrifying
visions of the future, I found practical applications of Donna Haraway's
"cyborg politics," the forging of new links between biotic elements and
technology (Haraway 1985). Organisms and machines had been joined
together to ground modest hopes, creating the possibility for a shared
future.

Reaching into tank 12 with gloved hands, turning over each leaf, the
scientist struggled to locate all of the frogs. This tank was home to half a

dozen lemur leaf frogs (*Agalychnis lemur*).[3] Lemur leaf frogs are nocturnal and spend daylight hours tucked up under leaves. At night, when the lemur leaf frogs are active, they are a reddish color, but during the day they assume a vibrant green. As she worked, some of the lemurs popped open their huge white eyes and began climbing toward the tank lid on spindly legs. The lemur tank was full of feces, as usual. She twisted the nozzle of the hose and began washing off the oblong blobs of mostly digested fruit flies through the wire mesh at the bottom of the tank.

In the wild, frogs hop away from their feces, reducing the risk of reinfecting themselves with diseases. Nematode worms and other parasites lay eggs in the digestive tracts of frogs and accumulate in the fecal pellets. Washing the feces away every day, and occasionally treating infected animals with drugs, helps maintain low parasite counts. Changing rubber gloves between tanks, or at least every time a glove contacts feces, protects the frogs from infecting each other. Every few days, the unbleached paper towels lining the bottom of the tanks are discarded and replaced. The trash stream of rubber gloves and paper towels is only one dimension of the costs incurred in keeping these endangered amphibians alive. Each month EVACC runs up an electricity bill of around $800 USD.

The EVACC facility is a fragile bubble of happiness sustained by a husband-and-wife team: Edgardo Griffith, a twenty-eight-year-old Panamanian biologist who often sports surfer's glasses, and Heidi Ross, an expatriate from the United States. It was created as a response to the growing sense of dread as a fungal disease spread across the highlands of Central America in a steady wave, about fifteen miles a year, driving scores of species to the brink of extinction. In March 2006, Edgardo "spotted a dead frog in a stream near El Valle. Its limbs were splayed out, and its skin was peeling. He scooped it up, went home and cried."[4] As the disease hit, Edgardo and Heidi set about collecting frogs and keeping them in conditions of strict biosecurity. They began working as *bricoleurs* and entrepreneurs, assembling networks of organisms and objects, making do with whatever beings and things were at hand. Hundreds of frogs took up temporary residence in a few vacant rooms of Hotel Campestre, a backpacker hotel in El Valle. Edgardo and Heidi began to cobble together everyday technologies into a life support system to protect frogs from the pathogenic fungus.

The Panamanian golden frog, *Atelopus zeteki*, quickly became a poster child for international conservation efforts. In any absolute sense, the

Figures 15.2, 15.3

A robust population of lemur frogs (*Hylomantis lemur*), a notably "cute" frog species, lives in the EVACC facility. Until recently this species was thought to be extinct in the neighboring country of Costa Rica, but then breeding populations were found on an abandoned farm and in a forested area near Barbilla National Park. The life cycle of the banded horned tree frog (*Hemiphractus fasciatus*, right) gives Rosalyn Diprose's (2002) notion of corporeal generosity a new twist. Eggs get pushed into a sack on the female's back as the male fertilizes them. These frogs breed by direct development, which means that when they are born, they pop out as small frogs instead of baby tadpoles. Then they stick around for a while, taking a ride on their mother's back. (Photographs by Brian Gratwicke and Edgardo Griffith. CC-BY-2.0, http://creativecommons.org/licenses/by/2.0, via Wikimedia Commons, http://fr.wikipedia.org/wiki/Agalychnis_lemur.)

Figure 15.4
Heidi Ross and Edgardo Griffith inside EVACC.

golden frog is not cuter than the lemur frog or the horned tree frog. But, as
this species reportedly went extinct in the wild in 2008, Edgardo and Heidi
became famous since they were keeping the last known populations alive
in Panama within their facility. Locally the couple became renowned for
their golden-frog-mobile, a four-wheel-drive jeep painted yellow with black
stripes. Golden frogs are featured on Panamanian lottery tickets and have
been scripted into stories about national patrimony and heritage. In conservation circles, golden frogs quickly became a flagship species—surrogates
that routinely stand in for other unloved frog species in fundraising campaigns. These charismatic animals captivated the imaginations of professional conservation biologists, volunteers, and donors, because they fit a

new and hopeful storyline. On the brink of extinction within their natal ecological communities, the golden frog has been saved by technological and scientific interventions.

Amid a demanding routine of daily care for the frogs living in Hotel Campestre, Edgardo and Heidi also began to navigate oblique powers structuring uneasy north–south relations (cf. García Canclini 2005). Wrangling with diverging values and obligations, they explored nonhierarchical modes of coexistence with large institutions. Major donors from North America—namely the Atlanta Botanical Gardens, Zoo Atlanta, and the Houston Zoo—began to lay the foundations for the EVACC buildings nearby, tailor-making a biosecure facility at a local zoo to replace their makeshift facility at Hotel Campestre. The Amphibian Ark, a transnational organization with a mission to "ensure the global survival of amphibians," became involved only after Edgardo and Heidi moved their frogs to the new building. The Ark was attempting to enroll facilities like EVACC into a global network of institutions "focusing on [species] that cannot currently be safeguarded in nature."[5] Visionaries at the helm of this Ark were producing hopes for endangered frogs at the intersection of concrete actions of care in the historical present and messianic dreams about a future to come (Rose, this vol.).

The Amphibian Ark dubbed 2008, the year I visited the EVACC facility, "the Year of the Frog." Kevin Zipple, the founder and principal leader of this global Ark, was aiming to raise a $50 million endowment to preserve endangered frog species in perpetuity. Upward of 3,900 species of amphibians, over one-half of all described frogs, salamanders, and caecilians, are in trouble according to the Ark's accounting.[6] In October 2007, as Zipple prepared to launch the Year of the Frog, he gave a speech at the Jackson Hole Wildlife Film Festival, where key gatekeepers for National Geographic, the Smithsonian, Discovery, Animal Planet, the journal *Nature*, and several other international media organizations were in attendance. His speech,

Figure 15.5
The logo for the Amphibian Ark's 2008 fundraising campaign.

showcasing the plight of the golden frog of Panama, opened with a pro-
vocative line: "Hi, I'm Kevin, and I'm building an ark." He continued:

"Amphibians are our modern day canaries in the coal mine," Zipple told the assembled VIPs.

"Just as the miners would take these sensitive birds with them into the mines,
and they would know if the birds died it was time to get out. Amphibians are raising
and waving red flags to us, saying: 'There is a serious problem, you need to change your
behavior, or you are going to suffer the same consequences.'

"A recent assessment of all amphibian species revealed that nearly half are declining. Somewhere between a third and a half are threatened with extinction. Just within in the past few decades well over 100 have already gone extinct. This is far more
severe than what we see with other vertebrate groups. And for every bird or mammal
species that is threatened with extinction, there are two to three amphibian species
that are on the verge."[7]

Kevin's rhetoric echoes Al Gore's language from *An Inconvenient Truth*, a
2006 documentary about global warming. This film has a secular apocalyptic narrative; it is a revoicing of environmental science in the language of
evangelical Christianity. These narratives use elements of the jeremiad, a
type of Protestant political sermon lamenting that people have fallen into
sinful ways and face ruin unless they swiftly reform. "Doom is imminent—
but conditional, not inevitable," Susan Friend Harding writes; "It can be
reversed by human action, but time is short" (Harding 2009). In Zipple's
speech, amphibians serve as sentinels, their fate foreshadowing that of
humans. Ongoing extinctions of frogs, salamanders, and caecilians prefig-
ure a possible future event in his imagination—the extinction of the human
species.

Jacques Derrida draws a helpful distinction between *apocalyptic* and *messianic* thinking (discussed in Jameson 1999, 63–64). Whereas apocalyptic
thinking looks toward absolute endings, messianic hopes, according to Derrida, contain "the attraction, invincible élan or affirmation of an unpredictable future-to-come (or even of a past-to-come-again)" (Derrida 1999,
253–254). "Not only must one not renounce the emancipatory desire," Derrida continues, "it is necessary to insist on it more than ever" (Derrida 1994,
74). As a figurehead at the helm of the Amphibian Ark, Kevin Zipple was
not just focused on definitive endings. As scores of frog species were going
extinct, he was trying to open up a moment of revolutionary time—a
moment when collective hopes about "saving the environment" or "preserving nature" might coalesce around the future of actual animals.

Derrida's hopeful sense of expectation is not oriented toward a specific Messiah. In contrast to Christian traditions, which pin hopes to the figure of Jesus Christ, Derrida's notion of messianicity is "without content" (Derrida 1994, 2004). Celebrating messianic desires that operate beyond the confines of any particular figure, he describes a universal structure of feeling that works independently of any specific historical moment or cultural location—a quasi-transcendental force he calls "messianicity without messianism" (Derrida 1999, 253). Derrida suggests that we should literally expect the unexpected by waiting for mysterious possibilities that are beyond our imaginative horizons (cf. Crapanzano 2004). Rather than pin hopes on something concrete, Derrida would have us wait for nothing in particular. Whereas the empty dreamscape of Derrida is haunted by a messianic spirit that refuses to be grounded in a particular figure, Kevin Zipple's imagination was focused on something specific: creating a livable future for a multitude of endangered animals. In other words, he was grounding modest biocultural hopes in hybrid assemblages of nature, culture, and technology (cf. Kirksey et al. 2014; Fortun 2001).

Messianic hopes in the biosciences are often problematic when they involve what Donna Haraway describes as "misplaced concreteness" (Haraway 1997b, 269). Biotech ventures have been criticized for using messianic discourse to focus the hopes of researchers, venture capitalists, and consumers on things that are too specific—like a gene, or a new pharmaceutical drug, or the resurrected cells of an extinct species (Fortun 2001; Sunder Rajan 2006). In terms of the Amphibian Ark, Zipple revoices speculative fictions and fabulations that have emerged at the intersection of biological sciences and economic enterprise to turn the rhetorical power of messianic discourse from producing profit for humans, to producing future generations of organisms he loves (Haraway 2014).

Awareness of the scale of the extinction wave sweeping through the worlds of amphibians led Zipple to recognize that isolated local efforts, like the EVACC facility in the highlands of Panama, were not capable of addressing the global crisis. Extinctions were taking place along a "dull edge" of time, in the words of Thom van Dooren, with a drawn-out and ongoing process of loss taking place long before and well after the final death (van Dooren 2014b). Against a prevailing sense of homogeneous, empty time—when nothing really seemed to change even amid definitive extinction events—Zipple was trying to open up revolutionary possibilities with his messianic language (see Benjamin 1968). He intended to galvanize the conservation community to raise an endowment to preserve frogs for eternity in biosecure breeding facilities and cryogenic banks. But the Year of the Frog, 2008, coincided with global financial disaster. Raising less than $1 million out of the $50 million goal of their capital campaign, the Amphibian Ark barely stayed afloat, scarcely covering their 2008 operating expenses.

Live Free or Die

Salvation can be an ambivalent grace. When animals are given the label "endangered species," according to Donna Haraway, they become subjected to the uncertain prospects of "being saved through a regulatory and technological apparatus of ecological and reproductive management" (Haraway 2014). While zoos and other breeding facilities style themselves as "salvific arks, bearing life's remnant and our hopes for redemption," this rhetoric often masks hidden regimes of violence (Chrulew 2011). As the messianic vision of Kevin Zipple largely failed to materialize—in the absence of an endowment capable of sustaining the life of endangered frog species in perpetuity—many committed caretakers like Edgardo Griffith and Heidi Ross soldiered on with limited resources, working to imagine and craft better futures for the animals in their facility. Others, with different political and ethical commitments, adopted more violent forms of care (see van Dooren 2014a,b).

After returning to the United States from Panama, I learned that another breeding population of golden frogs had been airlifted out of the country in 1999. The US military occupation of Panama, which had lasted nearly one hundred years, ended this same year. Scores of golden frogs were collected for a captive breeding program in the United States that continued a long history of unilateral action. The breeding program was aimed at "conserving genetic variability and maintaining viable captive populations." The Maryland Zoo in Baltimore was given an import permit, in accordance with the Convention for the International Trade in Endangered Species (CITES). This permit granted "ownership of the animals" to the zoo, rather than the Republic of Panama.[8]

After a few false starts, the Maryland Zoo enjoyed success in breeding golden frogs. Perhaps they were too successful. Every time breeding pairs mated they produced some 200–900 white eggs. Initially the biologists overseeing this conservation program were delighted at the fecundity of these animals. They were happy to see frogs flourishing within the artificial ecosystems they created. Soon, however, the Maryland Zoo ran out of space. They began shipping frogs around the United States—in plastic Gladware deli cups lined with damp toilet paper—to other zoos. These frogs are now common features of reptile houses. They are on display at institutions throughout North America: the Bronx Zoo, the Smithsonian's National Zoological Park in Washington, DC, the Atlanta Botanical Garden, the Toronto Zoo, and Busch Gardens in Tampa, Florida.[9]

As the zoological community began to run out of space, zookeepers in Baltimore started killing Panamanian golden frogs by the hundreds. Even after this species was presumed extinct in the wild, zoos culled their captive populations—selecting the most "genetically valuable" individuals to live. In July 2012 at "Herp Happy Hour" in Washington, DC, a monthly meetup of reptile and amphibian experts, one zookeeper told me: "Every time we have a new clutch of golden frogs I have to select sixty of the healthiest frogs to live. I can't stand the job of killing an endangered species, so I make my boss come in and euthanize the ones I don't select."[10] Writing of related dilemmas among bird conservationists, Thom van Dooren suggests that "many of us would still choose the violence of a conservation grounded in captive breeding over that of extinction" (van Dooren 2014b). Despite this, he insists that we "cannot be allowed to erase the genuine ethical difficulties," and that we should "consciously dwell within them, in an effort to, wherever possible, work towards something better" (ibid.).

Writing of interspecies love in the age of extinction, Deborah Bird Rose has argued for an ethics of care that does not exclude the possibility of death. "An ethical response to the call of others does not hinge on killing or not killing," she argues (Rose 2011, 18). Rather, the question becomes: What constitutes a good death? One prominent frog biologist who was also at the Herp Happy Hour in Washington told me that a "good death" cannot come from euthanasia at the hands of a zookeeper. Amid a sedate and melancholic conversation about biodiversity loss, financial woes, and zoo overcrowding, she suddenly slammed down her glass, spilling margarita on the table. Lifting her hand in a parody of a revolutionary salute, she shouted: "Live free or die!"

A deadly fungal disease is still present in the highlands of Panama. If the golden frogs were reintroduced to Panama and released, most would probably die. But by the reckoning of this researcher, a "good death" in the wild, connected to the hopes of adaptation and survival, is better than a "bad death" at the hands of a zookeeper. Despite hopes that some robust frogs might live if released, influential members of the conservation community are reluctant to let them go.

Michel Foucault understood the modern zoological garden as "a sort of happy, universalizing heterotopia" where "several spaces, several sites that are in themselves incompatible" are juxtaposed in a single real space (Foucault 1986). Zoos are indeed cosmopolitan collections of animals. As such, they are also breeding grounds for diseases from diverse corners of the world. Recent findings by veterinary pathologists suggest that zoos might be better understood as heterotopic hotbeds of parasitic protozoa, fungi, viruses, and bacteria. The officials of the Maryland Zoo are thus reluctant to send thousands of golden frogs back to Panama when they might not only succumb to the known fungal disease, but also inadvertently spread new amphibian diseases picked up during their stay in US zoological collections.

The Frog Fridge

My ethnographic methods involved volunteering time to care for frogs that had gone extinct in the wild. As a participant observer at facilities associated with the Amphibian Ark in Panama, as well as the Bronx Zoo in New York City, I cleaned cages, prepared food, and fed animals alongside zookeepers who were overworked and underpaid. While observing the toll of mind-numbing routines on human laborers and the cramped conditions for tens of thousands of animals living within a regime of institutionalized care, I crafted a concrete proposal for doing things differently. Grayson Earle, Mike Khadavi, and I created our collaborative artwork—*The Utopia for the Golden Frog of Panama*—in hopes of saving a few animals from euthanasia. We created our own biosecure holding tank. We adapted and used technologies that were ready-at-hand: household appliances, cheap digital hardware, and some specialized equipment from pet stores (cf. de Certeau 1998; da Costa and Philip 2008, xvii).

Our utopia was housed in an unused refrigerator enhanced with custom digital equipment, an aquarium, and a living ecosystem. This installation was our best attempt to interpret the interests and needs of another species. Hacking into the refrigerator with a power saw, we put a glass window in the front door. Grayson Earle also hacked into the electrical system of the refrigerator, creating a digital thermostat, using an Arduino, a small programmable microcontroller, to keep the fridge within 68–73°F daily, the ideal thermal range for golden frogs. Even as other kinds of animals were on the brink of extinction in polar regions, we retrofit one cooling machine as a cryopolitical proposal (see Radin and Kowal, this vol.).

We installed this artwork in Proteus Gowanus, an interdisciplinary gallery and reading room in Brooklyn. Golden frogs, a species endemic to cool highland climes of Central America, needed the retrofit refrigerator to survive the hot New York City summer. The fridge also provided resident frogs with an added layer of protection from the Gowanus Canal, a superfund site just outside the gallery that was laden with industrial toxins.[11] This installation at Proteus Gowanus was part of The Multispecies Salon, an exhibit that leveraged partnerships between artists and ethnographers to explore a set of interrelated questions:

Which beings flourish, and which fail, when natural and cultural worlds intermingle and collide? What happens when the bodies of organisms, and even entire ecosystems, are enlisted in the schemes of biotechnology and the dreams of biocapitalism? And finally, in the aftermath of disasters—in blasted landscapes that have been transformed by multiple catastrophes—what are the possibilities of biocultural hope? (Kirksey 2014)

We speculated that frogs living in sterile tanks, sitting day after day on a damp paper towel might—isolated from other species and companions that make forests livable and lively places—experience a sense of cosmic loneliness. As a partial solution to this problem, Mike Khadavi assembled a miniature ecosystem inside of the refrigerator with useful mosses and vascular plants collected from diverse corners of the globe. These plants were capable of generating enough oxygen to keep a small population of frogs alive. This feature, which meant that the frog fridge rarely needed to be opened, arguably made it more biosecure than the large pods of the Amphibian Ark, where humans were constantly coming and going, where frogs were living cheek-and-jowl with other amphibians in a heterotopic hotbed of disease. We also created a special composting system inside the tank and seeded

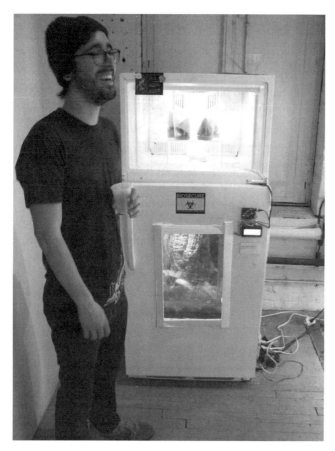

Figure 15.6
Grayson Earle pictured next to *The Utopia for the Golden Frog of Panama* at the Multi-species Salon in Brooklyn. Photograph by Eben Kirksey.

it with wingless *Drosophila melanogaster* and flightless *Drosophila hydei* mutants, which we ordered from an online retailer (Ed's Fly Meat, flymeat. com).[12] While we made no pretense of establishing conditions of sterility—conditions that are no more achievable in zoos or Amphibian Ark facilities—we carefully selected other species that have been demonstrated to be good for frogs to live with in multispecies worlds.

Aside from occasionally adding human food waste to the composting system, to generate future generations of fruit flies, *The Utopia for the Golden Frog* was built to function in relative autonomy—as long as it was plugged in to an electrical outlet. We mounted a webcam inside for live viewing, and posted a digital archive of temperature and humidity readings, enabling anyone who was interested to verify we had met the technical requirements to sustain the life of this species. In other words, we exposed our micro-biopolitical intervention to dominant regimes of biopolitics—opening up the frog fridge to regimes of surveillance (cf. Paxson 2014). Rather than Jeremy Bentham's Panopticon, where there might or might not be someone observing from a figural guard tower, we worked to recreate Bruno Latour's Oligopticon—where technologies of surveillance would enable a small group of people to accurately monitor an object of interest (Foucault and Sheridan 1991, 195; Latour 2005, 181).

After assembling *The Utopia for the Golden Frog*, and installing it for the opening of the Multispecies Salon in Brooklyn, I sent an email to the Maryland Zoo in Baltimore requesting frogs to populate the habitat. Kevin Murphy, Assistant Curator at the Zoo, was the "stud book holder" for the golden frog, a designation by the Association of Zoos and Aquariums (AZA) for the person who "dynamically documents the pedigree and entire demographic history of each individual in a population of a species."[13] In my email to Murphy, I asked how "I might submit a formal application to borrow some *Atelopus zeteki* adults for a temporary artistic display about the amphibian mass extinction crisis." I outlined the technical specifications of the frog fridge, adding: "The tank does not have any running water, so based on what I have learned about *Atelopus* reproductive biology I trust that this means the frogs won't be trying to breed."

Murphy wrote a friendly but dismissive note back, saying:

Unfortunately we are not able to provide golden frogs to non-AZA institutions. The original permitting agreement with the United States Fish and Wildlife Service (US-FWS) specifically prevents us from doing so. When we provide frogs to Universities for research the proposal has to clear The Zoo's Institutional Animal Care and Use Committee and we need to receive authorization from USFWS. Golden frogs are an endangered species and quite possibly functionally extinct in the wild so they are pretty heavily regulated.[14]

In parallel to this exchange with Murphy, I lobbied influential movers and shakers in amphibian worlds. But, I ultimately failed to convince the Maryland Zoo and the US Fish and Wildlife Service that a few frogs should be saved from euthanasia and kept in our modified, non-AZA affiliated refrigerator.

Capture and Escape

Kevin Murphy traveled to Panama in November 2013 to take part in a five-day workshop about the future of the Panamanian golden frog. Twenty-seven research scientists, conservationists, zookeepers, and government officials convened in El Valle, the site of Edgardo and Heidi's EVACC facility. They worked through tangled thickets of ethical, logistical, political, and epidemiological issues that plague conservation initiatives across national borders. Passions became inflamed, according to one attendee, when the subject of repatriating the golden frogs currently living in the United States was discussed. No one brought up the uncomfortable subject of euthanizing frogs, or the long legacy of unilateral action by US agents in the region, but the Panamanian delegation did repeatedly insist that their North American counterparts urgently needed to start finding new creative solutions to the problems at hand.[15]

The US zoological community had enfolded the golden frog into an emergent ecosystem—a system of holding tanks, public displays, and revenue-generating visitor attractions—that shows promise of sustaining itself into the future. The frogs had become entangled in a complex network of institutional alliances, bound to agents who had a vested interest in the status quo. Kevin Murphy and other US conservationists who visited El Valle in November 2013 were having difficulty imagining a time when they would have to let go. Thousands of golden frogs were being held fast by agents and institutions with competing values and obligations. These animals were entangled in relations of reciprocal capture. "In the case of symbiosis," writes Isabelle Stengers, reciprocal capture "is found to be positive: each of the beings coinvented by the relationship has an interest … in seeing the other maintain its existence" (Stengers 2010, 35–36). The case of golden frogs and US zoos is an example of reciprocal capture where the relationship is not entirely positive. I could only find hope in this contingent relationship if it contained the possibility that the frogs might one day escape.

The Maryland Zoo, the legal "owner" of the golden frogs, was entrusted by the US Fish and Wildlife Service to guard against this very possibility—that the frogs could escape not only their artificial zoo-based habitat, but also to escape regulation. The fear of the conservation officials, Kevin Zipple intimated, was that people—perhaps those involved with the market for rare and exotic animals—would collect any remaining golden frogs in the wilds of Panama, some of the most "genetically valuable" individuals in existence, and then launder them as animals bred in captivity. A small population of golden frogs was, in fact, recently rediscovered in Panama. The hopes of conservationists, who once presumed that this species was extinct in the wild, have been placed on the living figures of these frogs.

When I told Zipple about my art intervention, *The Utopia for the Golden Frog of Panama*, I descried it as modest proposal for letting citizen scientists participate in the effort to care for endangered species. Working with his own words, I said: "It seems like any way of expanding the carrying capacity of the Ark would be a good thing." "I agree," Zipple responded, before adding, "But, a lot of people in the private sector are motivated by the value of these animals. There is a tendency to sell them on the black market. There are just so many complicating factors to citizen involvement."

People who trade in endangered species, who collect them in the wild and sell them as pets, are certainly one of the problems leading to extinction. The pet trade for reptiles and amphibians is a booming underground economy in the United States. Millions of animals are imported each year from tropical forests (Collard 2014). Venturing into the realm of pet dealers, I attended major annual events—like the Daytona Reptile Breeders Expo in Florida and Frog Day in New York City. I also visited some of the premier retail establishments in the United States, Fauna on the Upper West Side of Manhattan and the East Bay Vivarium in Northern California.

Owen Maercks, who manages and owns the East Bay Vivarium, started off our interview by bluntly saying, "Many people in the pet trade are scumbags." Pointing me to a popular work of investigative journalism that exposed endangered species smuggling rings, *The Lizard King* by Bryan Christy (2008), he said, "I know all the people in this book. Don't think less of me. My main goal is to breed reptiles and frogs and sell them as pets." But Maercks also echoed the rhetoric of Kevin Zipple's Amphibian Ark: "When a forest is about to be turned into a farm or logged, we have locals running ahead of bulldozers, culling everything out of acreage. Catching stuff in the wild can be an act of salvation."

While scrutinizing some of the businesses involved in the international trade of amphibians and reptiles, I found disquieting examples of loosely regulated enterprises making money off of the reproductive capacities of animals. At the Daytona Expo I interviewed breeders who were using

Figure 15.7
Thousands of leopard geckos (*Eublepharis macularius*) and other animals were offered for sale at the Daytona Reptile Breeders Expo in small plastic containers. Photograph by Eben Kirksey.

do-it-yourself genetics to create "designer snakes" with customized color patterns. Chasing after the elusive genes that encode for pixilated configurations of color, and dreamy calico patterns, they were banking thousands of dollars on the possibility of producing an animal that would become a new unique object of desire. One legendary snake that sold for $100,000 was created by Trooper Walsh, who describes himself in his Facebook profile as "a Government Subsidized Snake and Dragon Farmer" whose favorite activity is "shooting guns." The characters populating the pet trade figure into the fears of Kevin Zipple, who worries that Panamanian golden frogs would be abused if they were to be adopted by citizen hobbyists instead of euthanized.

On the sidelines of big conventions orbiting around the sale of exotic and designer snakes, I also spoke with many frog enthusiasts—teachers, children, self-described "computer geeks." A divide separated the social worlds of people who were raising frogs and those who were breeding snakes. Prized frogs, on offer for prices ranging from $40 to $75, were being sold along with high-end vivariums—living ecosystems on par with the one created for my *Utopia* by Mike Khadavi, designed with the well-being of amphibians in mind. Following the frogs to people's homes, I found dedicated caretakers who were attentive to the needs of their pets and curious about their interests. Trolling through classified ads on Craigslist and other online forums, I came across a website dedicated to the trade of poison dart frogs (*Dendrobates*), which originate from Panama and Costa Rica. Dendroboard.com, "your source for dart frog information," is a community-run website where enthusiasts buy, sell, and trade animals. Self-taught experts also trade tips about animal care, aquarium construction, and veterinary concerns. The posts on Dendroboard reveal a lively virtual community where the genetic integrity of populations, the welfare of individual animals and other ethical issues are central concerns.

While I could not find any golden frogs for sale on the Internet, I found other rare and endangered species being bought and sold as pets. For example, the blue-sided tree frog (*Agalychnis annae*), a relative of the lemur frog from Costa Rica, has been proliferating in the pet trade in North America and Europe. Within Costa Rica the population of this frog species has declined by 50 percent since the 1990s. Conservationists believe that this sharp decline happened as a result of rampant fungal disease, predation on tadpoles by an introduced fish species, and the collection of animals in the wild for sale in the international pet trade. In 2007, the United States alone was reported to have imported 221,960 frogs belonging to the *Agalychnis* genus over the previous decade.[16] Once the blue-sided tree frog was formally designated as endangered by the IUCN Red List, the international trade of these animals was banned. Hobbyists in the United States and Europe continued to breed and sell these frogs, even after the ban.

From Germany, Martin Huber posted pictures of hundreds of tadpoles hatching in large plastic vats in his home, giving fellow frog enthusiasts daily updates on Frogforum.net in August 2011. Vendors in the United States were offering adult blue-sided tree frogs for sale at $20 each (figures 15.8, 15.9). Scientists in England were meanwhile breeding blue-sided tree frogs with a closely related species, *Agalychnis moreletii*, to produce unique hybrid offspring (Gray 2011).

Kevin Zipple's Amphibian Ark does not have enough carrying capacity to house blue-sided tree frogs. Against the backdrop of a grim political and

Figures 15.8, 15.9
The blue-sided tree frog has undergone a dramatic decline in protected parks in the highlands of Costa Rica. Currently populations are flourishing in polluted streams around San José, Costa Rica's capital, and also in the international pet trade. Pictures by Martin Huber (15.8) and Seth Kiser (15.9).

economic situation, only fifty species have found a home in his distributed ark, living at institutions that might be able to sustain long-term care.[17] This means that some 3,850 species of frogs with declining populations will not be saved. The Amphibian Ark is fragile, needy of care. Although troubling, the international pet trade may bring animals that have been orphaned by changing ecosystems into homes where they might receive care. Turning animals into commodities for exchange certainly has been one force contributing to the extinction of species. Even so, leveraging the economic value of endangered frogs, despite Zipple's concerns, might be the best way to actualize his conservation goals.

Exchange value and use value were classically part of Karl Marx's account of capital extraction and accumulation. Donna Haraway has added a surprising twist to Marx's classic story of capital with the notion of trans-species "encounter value." Lively capital, by Haraway's reckoning, can generate hopeful coalescences where "commerce and consciousness, evolution and bioengineering, and ethics and utilities are all at play" (Haraway 2008, 45–47). Lively capital is certainly at play in the social world of frog enthusiasts, who breed and feed animals from far-off lands in their own homes. Ethical forms of capitalist enterprise could thus help save the day in an era of mass extinction, when thousands of frog species are living in precarious situations.

Cryopolitics in financially strapped zoos means preserving life in a frozen form—trying to maintain natural forms in an unadulterated state, within complex articulations of technology and culture. Yet, other novel niches have emerged in landscapes that have been transformed by humans. Adopting a livelier approach might involve letting a multitude of people care for endangered frogs in their own homes—in utopias similar to the one that we designed but were unsuccessful in populating. Endangered frogs would then have the opportunity to invade and occupy bubbles of comfort created by people, to generate lively futures for themselves in air-conditioned living rooms, basements, and bedrooms throughout the industrialized world.

Future Promise

While conservation practitioners are crafting piecemeal solutions to reckon with the fact that many species of frogs can no longer live without protective infrastructures, some scientists are working at the frontiers of their imaginative horizons, searching for a breakthrough cure for the deadly fungal disease that is killing them in the wild. The Smithsonian National Zoo announced in 2012, via a blog post, that it has been inoculating frogs with experimental probiotic treatments in Front Royal, Virginia. "We usually think of bacteria as bad for us, but that isn't always the case," heralded the website: "For us humans, the most common examples of helpful bacteria, or probiotics, live in yogurt" (Smithsonian National Zoo 2012). The blog described how the redback salamander, a native of the Eastern United States, had survived the epidemic fungal outbreak. These salamanders had a diverse array of microbes living on their skin. Researchers speculated that multispecies communities could be a probiotic shield guarding against infection—a thin living bubble of protection. They hoped to discover microbes that could become new companions for endangered frogs, capable of producing chemical compounds with antibiotic (or at least antifungal) properties.[18]

Researchers placed their hopes on one kind of bacteria, *Janthinobacterium lividum*, which produced an antifungal compound called violacein in the lively microbiome on the skin of redback salamanders. When transferred to the mountain yellow-legged frog, an endangered species from California, these bacteria protected the frogs against the fungal pathogen. Captive Panamanian golden frogs, dubbed by the blog as "the poster-child for amphibian conservation," were also inoculated with probiotic bacteria treatments along with pathogenic fungi in experimental trials. While *Janthinobacterium lividum* bacteria initially kept fungal infections on golden frogs to a low level, eventually the probiotic microbes decreased in abundance and the frogs died (Bletz et al. 2013, 813; Baitchman and Pessier 2013, 680). In the face of this failure, I found cautious hopes proliferating in some scientific circles, searching through microbial worlds, coalescing around specific researchers, only to quickly dance away again in other directions. In Panama, I talked to researchers who were carefully testing some 600 kinds of microbes from the skin of other frogs—living figures of hope that might one day enable a multitude of golden frogs to live again in the wild.[19]

Speculation about scientific breakthroughs often fuels messianic thought in the biosciences (Sunder Rajan 2006).[20] "There can be no science without speculation," writes Mike Fortun; "there can be no economy without hype, there can be no 'now' without a contingent, promised, spectral and speculated future" (Fortun 2001). If messianic speculation in biology is often articulated through money-making dreams and schemes, different political, economic, and ethical forces are at play in hopes pinned on microbes that might help protect endangered amphibians (Franklin 2003). "Utopias, of course, do not last," in the words of Vicente Rafael, "even if their occasional and unexpected happenings are never the last" (Rafael 2003, 422).

Caretakers who maintain fragile forms of life within the Amphibian Ark are laboring in the present, working with scarce resources and following mind-numbing routines in the "now," while harboring dreams of a future that will be transformed by a scientific breakthrough, a silver bullet cure. While holding delicate animals inside uncomfortable architectures of incarceration, many conservation biologists are not jealously guarding their animals. Instead they are imagining a moment when their fragile bubbles might be broken open and the creatures in their care can escape.

Notes

1. The Amphibian Ark, "About Us: Activities," http://www.amphibianark.org/about-us/aark-activities/.

2. Cook, "2010: A Frog Odyssey," posted May 1, 2010, http://pinktreefrog.typepad.com/.

3. Solís et al., "*Hylomantis lemur*," posted January 1, 2008, http://www.iucnredlist.org/.

4. "Scientists Leap to Save Golden Frog in Panama," *Washington Post*, November 7, 2006.

5. The Amphibian Ark, "About Us: Activities," http://www.amphibianark.org/about-us/aark-activities/.

6. "About 30% (1,895) of the 6,285 amphibian species assessed by the IUCN are threatened with extinction," according to the Amphibian Ark website. "There are 6% (382) known to be Near Threatened and 25% (1,597) are data deficient. This means that about 3,900 species are in trouble." http://www.amphibianark.org/the-crisis/frightening-statistics/ (accessed August 5, 2014).

7. An edited video of this speech by Kevin Zipple has been posted YouTube. Amphibian Ark, "Our Planet's Canaries in the Coal Mines," posted July 25, 2011, http://www.youtube.com/.

8. Poole, "Husbandry Manual Panamanian Golden Frog," 3–4, www.ranadorada .org.

9. The Panamanian Golden Frogs were technically placed "on loan" from the Maryland Zoo. Only institutions in the Association of Zoos and Aquariums (AZA) were eligible to participate in the lending program, according to guidelines established by the US Fish and Wildlife Service. "This restriction was intended to prevent the protected species from entering the pet trade via captive zoo breeding," writes Vicky Poole, who authored the *Husbandry Manual for the Panamanian Golden Frog* while working for the National Aquarium in Baltimore. "Wild-caught illegal specimens could be 'laundered,'" according to Poole, "under the guise of coming from legal 'zoo stock.'" I conducted a series of interviews with Poole and also consulted her husbandry manual, available at www.ranadorada.org.

10. "Long-term captive management plan is to maintain 30–50 frogs from each bloodline," according to the Poole's husbandry manual. Frogs from some mating pairs "will be undesirables and may displace other more valuable offspring from a desirable breeding, so euthanasia will be necessary to eliminate, or at least reduce their numbers. Be prepared to house offspring indefinitely if allowed to survive." Poole, *Husbandry Manual Panamanian Golden Frog*, 13.

11. Superfund sites were established throughout the United States after the establishment of a government program to deal with places heavily contaminated by hazardous materials and requiring long-term rehabilitation.

12. Ordinary citizens who keep frogs and reptiles as pets now have ready access to the same genetically modified animals that have long been used by zoos and specialized breeding facilities. Mutan fruitflies are proliferating in biotechnical worlds and emergent ecologies orbiting around the pet industry. See Kohler 1994, 45.

13. Association of Zoos and Aquariums, "Studbooks," http://www.aza.org/studbooks (accessed February 17, 2014).

14. Email from Murphy to Kirksey, "Re: *Atelopus zeteki*," Tue., May 1, 2012, at 8:03 a.m.

15. Anonymous interview conducted by the author, Panama, February 17, 2014.

16. "Blue-sided tree frog (*Agalychnis annae*)," http://www.arkive.org/ (accessed June 1, 2014).

17. The Amphibian Ark, "Frightening Statistics," http://www.amphibianark.org (accessed March 3, 2014).

18. The Amphibian Ark, "Chytridiomycosis," http://www.amphibianark.org (accessed March 3, 2014).

19. In February 2014 I interviewed Myra Hughey at the Smithsonian Tropical Research Institute, who was collaborating with Reid Harris at James Madison University. Their work was still in the early stages at that time.

20. "Biotechnology occupies a messianic space," writes Sunder Rajan, "of technology and of Life linked through capital." Hopes of human patients, who are waiting for new cures, often become entangled with speculation by entrepreneurs who dream about cashing in with new miracle treatments. Money making schemes can nonetheless bring interesting things to life. Sunder Rajan 2006, 123, 149.

References

Amin, Ash. 2013. *Arts of the Political: New Openings for the Left*. Durham, NC: Duke University Press Books.

Amphibian Ark. n.d. About us: Activities. *Amphibian Ark*. http://www.amphibianark .org/about-us/aark-activities/.

Amphibian Ark. n.d. Chytridiomycosis. *Amphibian Ark*. http://www.amphibianark .org/the-crisis/frightening-statistics/ (accessed March 3, 2014).

Amphibian Ark. n.d. Frightening statistics. *Amphibian Ark*. http://www .amphibianark.org/the-crisis/frightening-statistics/ (accessed August 5, 2014).

Amphibian Ark Channel. 2011. *Our Planet's Canaries in the Coal Mines*. http://www .youtube.com/watch?v=xz0A7mTbz20&feature=youtube_gdata_player.

Association of Zoos and Aquariums. 2012. Studbooks. https://www.aza.org/ studbooks/.

Baitchman, Eric J., and Allan P. Pessier. 2013. Pathogenesis, Diagnosis, and Treatment of Amphibian Chytridiomycosis. *Veterinary Clinics of North America: Exotic Animal Practice* (Select Topics in Dermatology) 16 (3): 669–685. doi:10.1016/j. cvex.2013.05.009.

Benjamin, Walter. 1968. *Illuminations*. New York: Knopf Doubleday.

Bletz, Molly C., Andrew H. Loudon, Matthew H. Becker, Sara C. Bell, Douglas C. Woodhams, Kevin P. C. Minbiole, and Reid N. Harris. 2013. Mitigating amphibian chytridiomycosis with bioaugmentation: Characteristics of effective probiotics and strategies for their selection and use. *Ecology Letters* 16 (6): 807–820.

Catts, Oron, and Ionat Zurr. 2008. The ethics of experiential engagement with the manipulation of life. In *Tactical Biopolitics: Art, Activism, and Technoscience*, ed. Beatriz da Costa and Kavita Philip, 131. Cambridge, MA: MIT Press.

Christy, Bryan. 2008. *The Lizard King: The True Crimes and Passions of the World's Greatest Reptile Smugglers*. New York: Twelve.

Chrulew, Matthew. 2011. Managing love and death at the zoo: The biopolitics of endangered species preservation. *Australian Humanities Review* 50 (May): 137–157.

Collard, Rosemary-Claire. 2014. Putting animals back together, taking commodities apart. *Annals of the Association of American Geographers* 104 (1): 151–165.

Cook, Lucy. 2010. 2010: A Frog Odyssey. *The Amphibian Avenger.* http://pinktreefrog.typepad.com/amphibianavenger/2010/05/2010-a-frog-odyssey.html.

Crapanzano, Vincent. 2004. *Imaginative Horizons: An Essay in Literary-Philosophical Anthropology.* Chicago: University of Chicago Press.

da Costa, Beatriz, and Kavita Philip. 2008. *Tactical Biopolitics: Art, Activism, and Technoscience.* Cambridge, MA: MIT Press.

de Certeau, Michel. 1998. *The Practice of Everyday Life.* Minneapolis: University of Minnesota Press.

Derrida, Jacques. 1994. *Specters of Marx: The State of the Debt, the Work of Mourning, and the New International.* New York: Routledge.

Derrida, Jacques. 1999. Marx and sons. In *Ghostly Demarcations: A Symposium on Jacques Derrida's Spectres of Marx*, ed. Michael Sprinker, 213–70. London: Verso.

Derrida, Jacques. 2004. For a justice to come. Interview by Lieven De Cauter. http://hydra.humanities.uci.edu/derrida/brussels.html (accessed September 26, 2014).

Diprose, Rosalyn. 2002. *Corporeal Generosity: On Giving with Nietzsche, Merleau-Ponty, and Levinas.* Albany, NY: SUNY Press.

Fortun, Michael. 2001. Mediated speculations in the genomics futures markets. *New Genetics & Society* 20 (2): 139–156.

Foucault, Michel. 1986. Of other spaces: Utopias and heterotopias. Trans. Jay Miskowiec. *Diacritics* 16:22–27.

Foucault, Michel, and Alan Sheridan. 1991. *Discipline and Punish: The Birth of the Prison.* London: Penguin.

Franklin, Sarah. 2003. Ethical biocapital: New strategies of cell culture. In *Remaking Life & Death: Toward an Anthropology of the Biosciences*, ed. Margaret Lock and Sarah Franklin. Santa Fe: Oxford: SAR Press.

García Canclini, Néstor. 2005. *Hybrid Cultures: Strategies for Entering and Leaving Modernity*, exp. ed. Trans. Christopher L. Chiappari and Silvia L. Lopez. Minneapolis: University of Minnesota Press.

Gray, Andrew R. 2011. Notes on hybridization in leaf frogs of the genus Agalychnis (Anura, Hylidae, Phyllomedusinae). http://arxiv.org/abs/1102.4039.

Haraway, Donna. 1985. A manifesto for cyborgs: Science, technology, and socialist feminism in the 80s. *Socialist Review* 15 (80): 65–107.

Haraway, Donna. 1997a. Speculative fabulations for technoculture's generations: Taking care of unexpected country. In *Catalogue of the ARTIUM Exhibition of Patricia Piccini's Art*, 100–107. Vitoria, Spain.

Haraway, Donna. 1997b. *Modest_Witness@Second_Millennium.FemaleMan_Meets _OncoMouse: Feminism and Technoscience. 1ST edition.* New York: Routledge.

Haraway, Donna. 2008. *When Species Meet.* Minneapolis: University of Minnesota Press.

Haraway, Donna. 2014. Speculative fabulations for technoculture's generations. In *The Multispecies Salon*, ed. Eben Kirksey. Durham, NC: Duke University Press Books.

Harding, Susan Friend. 2009. Get religion. In *The Insecure American: How We Got Here and What We Should Do about It*, ed. Hugh Gusterson and Catherine Besteman, 345–361. Berkeley: University of California Press.

Jameson, Frederic. 1999. Marx's purloined letter. In *Ghostly Demarcations: A Symposium on Jacques Derrida's Spectres of Marx*, ed. Michael Sprinker. London: Verso.

Kirksey, Eben. 2015. *Emergent Ecologies.* Durham, NC: Duke University Press.

Kirksey, Eben. 2014. *The Multispecies Salon.* Durham, NC: Duke University Press Books.

Kirksey, S. Eben, Nicholas Shapiro, and Maria Brodine. 2014. Hope in blasted landscapes. In *The Multispecies Salon*, ed. Eben Kirksey. Durham, NC: Duke University Press Books.

Kohler, Robert E. 1994. *Lords of the Fly: Drosophila Genetics and the Experimental Life.* Chicago: University of Chicago Press.

Latour, Bruno. 2005. *Reassembling the Social: An Introduction to Actor-Network-Theory.* Oxford: Oxford University Press.

Paxson, Heather. 2014. Microbiopolitics. In *The Multispecies Salon*, ed. Eben Kirksey. Durham, NC: Duke University Press Books.

Poole, Vicky. 2006. *Husbandry Manual Panamanian Golden Frog.* Project Golden Frog Proyecto Rana Dorada. http://www.ranadorada.org/PDF/HusbandryManual.pdf.

Rafael, Vicente L. 2003. The cell phone and the crowd: Messianic politics in the contemporary Philippines. *Public Culture* 15 (3): 399–425.

Rose, Deborah. 2011. *Wild Dog Dreaming: Love and Extinction.* Charlottesville: University of Virginia Press.

Rose, Deborah. 2014. Flying foxes: Kin, keystone, kontaminant. http://www.academia.edu/4539623/Flying_Foxes_Kin_Keystone_Kontaminant. (accessed August 3, 2014).

Scientists leap to save golden frog in Panama. 2006. *Washington Post*, November 7.

Smithsonian National Zoo. 2012. New Experiment May Offer Hope for Frogs Facing Chytrid. *Amphibian Rescue and Conservation Project*. http://amphibianrescue.org/2012/05/28/new-experiment-may-offer-hope-for-frogs-facing-chytrid/.

Solis, F., et al. 2008. Agalychnis lemur. *ICUN Red List of Threatened Species*. http://www.iucnredlist.org/details/55855/0.

Stengers, Isabelle. 2010. *Cosmopolitics I*. Trans. R. Bononno. Minneapolis: University of Minnesota Press.

Stengers, Isabelle. 2011. *Cosmopolitics II*. Minneapolis: University of Minnesota Press.

Sunder Rajan, Kaushik. 2006. *Biocapital: The Constitution of Postgenomic Life*. Durham, NC: Duke University Press.

van Dooren, Thom. 2014a. Authentic crows: Identity, captivity, and endangered species conservation. http://www.academia.edu/5759237/Authentic_crows_Identity_captivity_and_endangered_species_conservation (accessed August 3, 2014).

van Dooren, Thom. 2014b. *Flight Ways: Life and Loss at the Edge of Extinction*. New York: Columbia University Press.

Wildscreen Arkive. n.d. Blue-sided tree frog (Agalychnis annae). http://www.arkive.org/blue-sided-tree-frog/agalychnis-annae/ (accessed June 1, 2014).

Williams, Raymond. 1978. Utopia and science fiction. *Science Fiction Studies* 5: 203–214.

16 The Cryopolitics of Survival from the Cold War to the Present: A Fugue

Charis Thompson

Freezing has played a fugue through my life.[1] This marks me out as someone who was born at the height of the Cold War, the era when the eponymous temperature was cold, not hot as it is for my children. This generation has always known global warming; I was raised on nuclear winter. The conflict they know is an uncontrolled burn; it has razed the bounds of the nation-state and fueled the collapse of a distinction between war and genocide, military and civilian. The Cold War, on the other hand, was all nations and blocs, treaties and nonalignment; politics by other means, war by other means.[2]

Not only am I child of the Cold War, I grew up in a United Nations and World Health Organization family, spending the first part of my childhood in what was referred to as the Tropics. Ice was a triumph and a never-ending challenge. The Empire had risen and fallen on ice; now the smallpox vaccine was stored, attenuated, on ice, and Western corporations and local businesses made and moved ice in quantities small and large (David 2011; Herold 2011; Radin 2013).

Growing up in that part of the world, we were warned not to eat ice. It was bound to be contaminated. But it kept ice cream cold and accompanied the ubiquitous soft drinks. It dripped from under the insulating sawdust as it moved through the streets, and melted quickly in plastic bags of sweet iced tea. In the expatriate household, it hardened into small freezer-burned cubes, riven with shards, in slightly battered aluminum trays in a top corner of the fridge. In the evening, it was banged out into cocktail glasses.

We returned to Europe—Geneva, headquarters of several UN agencies—and the refrigerator still broke down. When the repairman came to the house, my mother—she was literary—wrote on the family calendar, "The Iceman Cometh."[3] He kept the middle classes in domestic ice. Americans, long-time ice innovators, got us to the moon first, and "iceman" became a synonym for a heartless killer.[4] We threw away a lot of overwrapped and overprocessed food that may or may not have spoiled. We didn't recycle any of our garbage, but the Swiss all had nuclear fallout shelters in their basements.

Becoming a mother as the Cold War was ending, I craved ice. I chomped through pregnancy, testing the tooth enamel that had come to fluoride slightly too late. This time, ice could be counted on to fight off nausea, not induce it. I poorly balanced work and personal life. Among my friends and especially my colleagues, childbearing was often postponed in the name of this elusive balance, and more and more women and men reproduced with the help of cryopreservation.

When I was working on my PhD, freezing was central to the means of reproduction, as I decided on my field of study (Thompson 2007, 1998, 1999). At the endangered animal field sites I worked at in East Africa and the United States, frozen zoos were mobilized to collapse phylogenetic and ontogenetic time, in the name of biodiversity conservation for the future (Lermen et al. 2009). When I worked in situ in East Africa, Coca Cola or Pepsi on ice showed up at every meeting, even when the school where the meeting on community-based conservation was taking place had no books (Thompson 2002).

In my assisted reproductive technology sites, freezing coordinated bodies with technology, and with one another across borders and in time. Freezing was the *sine qua non* of in vitro, and of its corollary, *ex vivo*. Gametes could move, thanks to freezing, without bodies, and between bodies, breaking the putative link with nature and biological reproduction of the nuclear family (Landecker 2007; Franklin 2013; Hoeyer 2013).

The first time I topped up a liquid nitrogen freezer that held reproductively priceless frozen human embryos, I didn't understand freezing substances well enough. I poured the liquid nitrogen as if it were beer, letting go of the depressed lever on the nozzle when the nitrogen appeared to be some centimeters below the canister's rim. I stepped back. The liquid nitrogen continued slowly to rise and rise, more yeast than beer itself, and began gradually to fill the small office as if it were dry ice.

Working in an ART clinic is a this-worldly and embodied experience, getting on with helping would-be reproductive adults initiate some form of biological family building. Amid the work to help get Ms. X and Ms. V, or Mr Z. and Mr. Y, or Mr. and Mrs. W, or single would-be mothers or fathers to the state of "expecting," one could lose sight of the wider cultural drama-turgy of reproductive technologies. A room full of billowing, overflowed liquid nitrogen cued the sci-fi B-movie; test-tube babies were still futuristic, the stuff of science fiction.

I then turned to regenerative medicine—or, rather, one of its basic sciences, pluripotent stem cell research, now sustained by leftover ART embryos (Thompson 2013). My interest in science architecture burgeoned. All life science lab benches used to have gas and air. All labs had hoods. All had animal facilities. Now data storage and management, and freezing and cooling facilities, were also being built into the very infrastructure of the new cathedrals to science by design (Thompson 2015). Data and biosam-ples both need cooling. In 2013, I toured the Francis Crick Institute in Lon-don while it was still a building site; being able to cool and freeze centrally had been made integral to its flexible and translational design for twenty-first-century life and biomedical science (Francis Crick Institute 2012).[5]

Geopolitical Cold was a system in constant ideological production in my day. The so-called "First," "Second," and "Third Worlds" were all enrolled. Boys became men; the Cold War affect was Trust, which no one had but everyone sought, especially around the technology that oxymoronically deterred through Mutually Assured Destruction, and everyone was a poten-tial double agent.[6] East/West masculinity drew the heat out of politics with its escalating weapons systems, while men from the Global South were dis-appeared in the imagery of failed states and famine that featured "orphaned" children and "native" women, ravaged physically by "Slim disease" and economically by structural adjustment and the other postcolonial realities and fictions of hard and soft power.[7]

Women were also used ideologically to dissipate and transform the heat of conflict into the realms of the domestic and civilian. Some of the more affluent and pale women in the United States, among those actually living in nuclear families, drugged themselves into domestic oblivion (Metzl 2003). Refrigerator mothers were accused of cathecting failures of warmth onto their children, producing psychiatric disorder (Kanner and Eisenberg 1956). From Hitchcock to James Bond, glamorous film stars aided and abetted Coldness through seductive courtship with undercover danger (Thompson 2003).

Women's empowerment in the Global South during the Cold War was co-optation on a transnational scale, handmaiden to a discourse of devel-opment, population control, and demographic transition (Murphy 2012;

Bandarage 1997; Halfon 2007). If girls would only go to school, and women take microcredit, and desire fewer children, developing countries could become developed without having to go through their own costly, long, and deadly industrial and capitalist revolutions—they could simply join the West/First World.[8]

Women's empowerment and trust and all the rest of it was "ideology." I hated and loved that word. It patrolled our struggles to keep ever-moving boundaries between the real and coerced, authentic and fake, known and covert critically alive amid the inexorable systems of East/West, North/South, racism, sexism, and physical and mental illness. Cold War, violence by ideology, war by other means, nurtured the hidden persuaders of capitalism, the theft of the legacies of Empire, the onslaught to free speech of McCarthyism, and the authoritarianism of Communism (Packard 1957). But as long as it was ideology, as long as it was Cold, I could imagine that it could be otherwise.

It is an "inconvenient truth" that my generation has bequeathed to my kids' generation a world that has already abandoned the ice.[9] Photos of polar bears cowering on floating blocks of broken-off ice caps are present-day icons of climate change, even if they are reminiscent to my generation of those Cold War Jell-O ads of a polar bear on a lump of ice proclaiming the "cool dessert."[10] Today, polar bears are losing their habitat and thus their food, and tragedies can happen (e.g., "Polar Bear Kills Young British Adventurer in Norway" 2011).

Women's empowerment is still going strong in civilizational ideology, but a warming world scripts the link between female gendering and world saving differently. Corporate America's blockbuster movie *Frozen* tells us that girl power is enough to re-create ice; they need only "Let It Go."[11] Furthermore, if they "Lean In," eschew foolish boys, and recognize a bit of sister love, they can learn to use their ice-making powers in ways that reconcile human civilization and a re-iced nature (see Sandberg 2013).[12]

The *Frozen* sisters—one a blonde ice queen who turns her destructive ice making superpower into a creative gift while shut off from love and humanity, and the other her hot-headed but equally Northern European redheaded sister—and their motley crew of humans and nonhumans know no Global South, and their far-north indigenous population has no sovereignty.[13] This is Nordic fairy tale turned neoliberal. All the reasons why the ice is disappearing are themselves disappeared in a remedy based on the fictive white- and pink-washed power of girls. It is a staple of mainstream US history that the Quakers "came to do good and ended up doing well."[14] In Disney's world, where girls reconcile climate change and the real world through sheer girl-power resilience, girls come to feel good and do good while looking good, and must keep on doing it all (Hinshaw and Kranz 2009). When Facebook and Apple offered their employees coverage for egg freezing, cryopreserved eggs became a similarly neoliberal way for women to postpone childbearing so as to get ahead, and a way for Silicon Valley to deal with its gender problem (Cussins 2014).

No doubt, freezing will continue its fugue-like presence in my life. But it will always be different for me than it is for my kids. It is not that the cold has gone away from their too warm world: they can judge what's hot and what's not *and* what's cool far better than I ever could. But I knew cold as the geopolitical temperature of the planet, when it was politics by other means to get ice circulating across cultures. My kids' generation knows cold as something oxymoronically made extreme by excess warming, and they live in a world chronically short of ice in nature, not in culture.

There may not be much to choose between a world that is imminently, disastrously cold and one that is imminently, disastrously warm, between the Iceman and the Ice maiden. In the summer of 2014, the world connected by Facebook dumped buckets of ice on its heads in a viral crowdsourced fundraiser for ALS disease, raising well over $100 million. It did not take long for the "ice challenge" to be hacked—to protest the waste of water and ice, and the underlying structural inequalities that contribute massively to health inequalities. There is a lesson here. Could women's empowerment be deployed outside of the logics of the US State Department and development agencies? Could girl power mitigate rather than obscure structural inequality, cooling current conflicts and the warming planet? Cryopolitics demands that we continue to take the temperature of the times, but also that we consider temperature over time and across bodies and borders. As we pay attention to hot and cold, and to those who make and lose and take and use ice, we can hope to keep our critical selves alive and well enough to fight for another day.

Notes

1. On the fugue, meaning, and metaphor, see Hofstadter 1979.

2. The idea that war is politics by other means is usually attributed to Carl von Clausewitz's *Vom Kriege* (Dummlers Verlag, 1832). I am here suggesting that both war and politics were sublimated in the other during the Cold War.

3. Eugene O'Neill's play, *The Iceman Cometh*, had premiered in 1946.

4. The contract and serial killer Richard Kuklinski (1935–2006), who murdered hundreds of people from 1948 to 1986, was known as the "Iceman" for freezing some of his victims in an industrial freezer to conceal the time of death.

5. An article describing the building's design concept says, "Shared support spaces—items like freezers, coolers, tissue culture rooms, opaque rooms—are in the center" (Francis Crick Institute 2012).

6. Mutually Assured Destruction, or MAD, was the acronym for the military/security strategy associated with the Cold War of deterrence through the prospect of attack by either side precipitating annihilation of both sides. On nuclearity, affect, and the links between nuclearity and climate change, see the work of Joseph Masco, especially Masco 2006, 2009.

7. "Slim disease" was the name given to the wasting phase of HIV/AIDS in East and Southern Africa in the 1980s, prior to the identification of the HIV virus. "Soft power" is a phrase attributed to Joseph Nye of the Kennedy School of Government at Harvard; he began using the term in the 1980s (Nye 2004).

8. For a statement and assessment over time of these concepts, see Pomeranz 2014.

9. *An Inconvenient Truth* was the 2006 film directed by Davis Guggenheim about US Vice President Al Gore's attempt to get people to acknowledge the threat of global climate change.

10. A 1954 Jell-O ad showed a polar bear on a small cushion of ice, proclaiming it a "cool dessert."

11. *Frozen* is a 2013 Disney film, adapted from Hans Christian Andersen's classic, *The Snow Queen*; "Let It Go" is a song from the film that achieved wide popularity.

12. Sheryl Sandberg, author of *Lean In*, is the operating officer of Facebook.

13. The stand-ins for indigenous people are nonhuman "rock trolls" that wear crystals around their rocky necks and are accompanied by African-themed music.

14. This formulation is used in PBS's online "History of Affluenza," at http://www.pbs.org/kcts/affluenza/diag/hist1.html.

References

Bandarage, Asoka. 1997. *Women, Population, and Global Crisis*. London: Zed Books.

Cussins, Jessica. 2014. Dear Facebook, please don't tell women to lean in to egg freezing. *Huffington Post*, October 16. http://www.huffingtonpost.com/jessica -cussins/dear-facebook-please-dont_b_5993646.html.

David, Elizabeth. 2011. *Harvest of the Cold Months: The Social History of Ice and Ices*. London: Faber Finds.

Francis Crick Institute. 2012. The Francis Crick Institute: Biomedical research without barriers. October 30. http://www.tradelineinc.com/reports/2012-10/francis-crick -institute-biomedical-research-without-barriers

Franklin, Sarah. 2013. *Biological Relatives: IVF, Stem Cells, and the Future of Kinship*. Durham, NC: Duke University Press.

Hoeyer, Klaus. 2013. *Exchanging Human Bodily Material: Rethinking Bodies and Markets*. New York: Springer.

Hinshaw, Stephen, and Rachel Kranz. 2009. *The Triple Bind: Saving Our Teenage Girls from Today's Pressures*. New York: Ballantine Books.

Halfon, Saul. 2007. *The Cairo Consensus: Demographic Surveys, Women's Empowerment, and Regime Change in Population Policy*. Lanham, MD: Lexington Books.

Herold, Marc. 2011. Tropics: The export of "crystal blocks of Yankee Coldness" to India and Brazil. *Revista Espaco Academico* 126:162–177.

Hofstadter, Douglas. 1979. *Gödel, Escher, Bach: An Eternal Golden Braid*. New York: Basic Books.

Kanner, L., and L. Eisenberg. 1956. Early infantile autism 1943–1955. *American Journal of Orthopsychiatry* 26 (3): 556–566.

Landecker, Hannah. 2007. *Culturing Life: How Cells Became Technologies*. Cambridge, MA: Harvard University Press.

Lermen, D., B. Blömeke, R. Browne, A. Clarke, P. W. Dyce, et al. 2009. Cryobanking of viable biomaterials: Implementation of new strategies for conservation purposes. *Molecular Ecology* 18 (6): 1030–1033.

Masco, Joseph. 2006. *The Nuclear Borderlands: The Manhattan Project and Post-Cold War New Mexico*. Princeton, NJ: Princeton University Press.

Masco, Joseph. 2009. Bad weather: On planetary crisis. *Social Studies of Science* 40 (1): 7–40.

Metzl, Jonathan. 2003. Mother's little helper: The crisis of psychoanalysis and the Miltown Resolution. *Gender & History* 15 (2): 228–255.

Murphy, Michelle. 2012. *Seizing the Means of Reproduction: Entanglements of Feminism, Health, and Technoscience*. Durham, NC: Duke University Press.

Nye, Joseph. 2004. *Soft Power: The Means to Success in World Politics*. New York: Public Affairs.

Packard, Vance. 1957. *The Hidden Persuaders*. New York: Ig Publishing.

Polar bear kills young British adventurer in Norway. 2011. *Guardian*, August 5. http://www.theguardian.com/world/2011/aug/05/polar-bear-mauls-british-death.

Pomeranz, Dina. 2014. *The Promise of Microfinance and Women's Empowerment*. Ey.com, February. http://www.ey.com/Publication/vwLUAssets/EY_-_Microfinance_and_womens_empowerment/$FILE/EY-The%20promise-of-microfinance-and-womens-empowerment.pdf.

Radin, Joanna. 2013. Latent Life: Concepts and Practices of Tissue Preservation in the International Biological Program. *Social Studies of Science* 43 (4): 483–508.

Sandberg, Sheryl. 2013. *Lean In: Women, Work, and the Will to Lead*. London: W. H. Allen.

Thompson, Charis. 1998. Quit sniveling cryo-baby, we'll work out which one's your mama: Kinship in an infertility clinic. In *Cyborg Babies: From Techno-Sex to Techno-Tots*, ed. Robbie Davis-Floyd and Joe Dumit, 40–66. Abingdon-on-Thames: Routledge.

Thompson, Charis. 1999. Confessions of a bioterrorist: Subject position and the valuing of reproductions. In *Playing Dolly: Technocultural Formations, Fantasies, and Fictions of Assisted Reproduction*, ed. E. Ann Kaplan and Susan Squier, 189–219. New Brunswick, NJ: Rutgers University Press.

Thompson, Charis. 2003. Comment: Forum on gender and nation in post-war visual culture. *Gender & History* 15:262–267.

Thompson, Charis. 2002. When elephants stand for competing models of nature. In *Complexities: Social Studies of Knowledge Practices*, ed. Annemarie Mol and John Law, 166–190. Durham, NC: Duke University Press.

Thompson, Charis. 2007. *Making Parents: The Ontological Choreography of Reproductive Technology*. Cambridge, MA: MIT Press.

Thompson, Charis. 2013. *Good Science: The Ethical Choreography of Stem Cell Research*. Cambridge, MA: MIT Press.

Thompson, Charis. 2015. Designing for the life sciences: The epistemology of elite life science real estate. *Tecnoscienza: Italian Journal of Science and Technology Studies* 5 (2): 13–44.

Contributors

Warwick Anderson is ARC Laureate Fellow and Professor in the Department of History and the Centre for Values, Ethics and the Law in Medicine, at the University of Sydney. Additionally, he is affiliated with the Unit for History and Philosophy of Science at Sydney and is a Professorial Fellow of the School of Population Health at the University of Melbourne. He is the author of many articles and books, the most recent of which is (with Ian Mackay) *Intolerant Bodies: A Short History of Autoimmunity* (2014).

Michael Bravo is University Senior Lecturer and Fellow of Downing College, University of Cambridge. He is Convenor of the Circumpolar History and Public Policy Research Group, Scott Polar Research Institute, Member of the Department of Geography's Natures, Cultures, and Knowledges Thematic Research Group, and is also a research associate in the Department of the History and Philosophy of Science. His most recent book is *Arctic Geopolitics and Autonomy* (2011), a collaboration with Nicola Triscott of Arts Catalyst (London) and the Slovenian artist Marko Peljhan.

Jonny Bunning is a PhD candidate in the Department of History of Science and Medicine at Yale University. He is working on a dissertation on the history of the global political economy.

Matthew Chrulew is a Research Fellow in the Centre for Culture and Technology at Curtin University, where he leads the Posthumanism and Technology research program. In addition to authoring many essays and editing many volumes in the realm of extinction studies, he is an Associate Editor of the journal *Environmental Humanities*.

Soraya de Chadarevian is Professor in the Department of History and the Institute for Society and Genetics at University of California, Los Angeles.

Among her many publications is the book *Designs for Life: Molecular Biology after World War II* (2002; repr. 2003).

Alexander Friedrich is an interdisciplinary researcher based at the Technische Universität Darmstadt. As a postdoctoral scholar of the research group Topology of Technology at Technische Universität Darmstadt he is currently working on a research project on refrigeration as a biopower.

Klaus Hoeyer is Professor in the Section of Health Services Research and the Department of Public Health at the University of Copenhagen. He has written extensively at the intersection of ethics, bodily materials, and health data. His most recent book is *Exchanging Human Bodily Material: Rethinking Bodies and Markets* (2013).

Frédéric Keck is an anthropologist and Researcher at the Laboratoire d'anthropologie Sociale and the Head of the Research Department of the Musée du Quai Branly. He has published many works on the history of social anthropology and has a forthcoming book on contemporary biosecurity practices.

Eben Kirksey is a DECRA Fellow of the Australian Research Council and Senior Lecturer in Environmental Humanities at the University of New South Wales. He is the author of numerous books and articles on multispecies ethnography. His most recent book is *Emergent Ecologies* (2015).

Emma Kowal is Professor of Anthropology in the School of Humanities and Social Sciences and the Alfred Deakin Institute for Citizenship and Globalisation at Deakin University. She is the author of *Trapped in the Gap: Doing Good in Indigenous Australia* (2015).

Joanna Radin is Assistant Professor of History of Science and Medicine at Yale University, where she also has appointments in the departments of History and Anthropology. She is the author of *Life on Ice: A History of New Uses for Cold Blood* (2017).

Deborah Bird Rose is an Adjunct Professor in Environmental Humanities and a member of the Extinction Studies Working Group at the University of New South Wales. She has many publications in eco-criticism, including the book *Wild Dog Dreaming: Love and Extinction* (2011).

Kim TallBear is Associate Professor of Native Studies at the University of Alberta. She is also a member of the Oak Lake Writers, a group of Dakota,

Lakota, and Nakota (Oceti Sakowin) writers. She is the author of *Native American DNA: Tribal Belonging and the False Promise of Genetic Science* (2013).

Charis Thompson is Professor of Sociology at the London School of Economics. She has published widely on the sociology of gender and biomedical technology. Her most recent book is *Good Science: The Ethical Choreography of Stem Cell Research* (2013).

David Turnbull is a cultural anthropologist and Senior Research Fellow at the Victorian Eco Innovation Lab of the Faculty of Architecture at University of Melbourne. Among his many publications at the intersection of indigenous and postcolonial theory is the book *Masons, Tricksters, and Cartographers* (2000).

Thom van Dooren is Senior Lecturer in Environmental Humanities at the University of New South Wales in Sydney, Australia, and coeditor of the journal *Environmental Humanities*. His most recent book is *Flight Ways: Life and Loss at the Edge of Extinction* (2014).

Rebecca J. H. Woods is Assistant Professor at the Institute for History and Philosophy of Science and Technology at the University of Toronto. She is completing a book, *The Herds Shot Round the World: Native Breeds and the British Empire, 1800–1900*, which will inaugurate a new environmental history series at the UNC Press.

Index